WILD YEAST BREAD MAKING

天然酵母麵包

Unique flavours, texture and aroma with every artisan bread you bake

不同風味・不同口感
每一個麵包都洋溢獨特芳香

Reko Sham

REKO

/作者簡介/

從一個牛油蛋糕愛上烘焙，

享受在家中小廚房做實驗；

愛吃，追求製作出美味精緻的食物；

愛攝影，把製成品凝在瞬間，記下快樂的片段。

烘焙的樂趣除可靜靜細味外，我們亦可以交流彼此的心得；

在培養天然酵母、烘焙麵包過程中的苦與樂、成與敗都可與我分享，

我的email: reko24@yahoo.com.hk

我的facebook: www.facebook.com/rekohomemade

/前言/

這次出書從構思、寫食譜、食物造型、拍攝，也是自己一手包辦。因為平日一直都是這樣做，所以過程開心、輕鬆，很享受。

對於烘焙和攝影，我不是專業人士，只因有強烈興趣使我著迷的去研究、去實踐。我算是完美主義者，對食物質素和外觀也忠於自己定立的要求，所以一直從失敗中學習，烘焙我滿意的麵包。

這次出書，是想把基礎的天然酵母製作，介紹給對自釀酵母感到困難的人士，也想把我實驗後喜歡的包點介紹給喜歡自家製的烘焙愛好者；想告訴大家，在家中製作高品質、美味精緻的麵包絕不是困難的事。

因為各地自家培育和製作天然酵母的方法，種類繁多，沉醉實驗使我每天充滿新挑戰。感恩在生活中找到了烹飪、烘焙、攝影這些興趣，使我做麵包也樂透半天，充實了我的生命。

/目錄/

尋找麵包真正的味道

由基本學起

牛角的變化

較複雜，但美味

在食譜內，我把焗爐預熱高於實際烘焙的度數，因為打開焗爐門時，焗爐內的溫度會急速下降，預熱時設定溫度高30℃，便可確保麵糰入爐時在適當溫度烘烤。

Variations of Croissant

Slightly more complicated, but well worth it

I usually preheat the oven 30°C higher than the actual baking temperature. It's because the oven is cooled down drastically once you open the door. When the oven is preheated to a higher temperature, you can make sure the dough is baked at the desired temperature right at the moment you put it in the oven.

/尋找麵包真正的味道/

使用全天然的材料培育自家製天然酵母，這不是複雜的事，反而製作方便，使用簡單，麵糰經過長時間低溫或冷藏發酵，緩慢發酵可把原料風味引發出來，麵糰也能完整地發展，促進熟成，發揮天然酵母菌的優點，把麵包真正的味道和風味呈現出來。

自家培育的天然酵母是野生酵母菌。野生酵母菌無處不在，可以用來製作天然酵母的食材也十分廣泛，製作酵母液的原料也種類繁多。

有許多種培育天然酵母的方法，以下介紹的是製作簡單、成功率高的基礎葡萄乾種和草莓酵種。以葡萄乾製成的酵母液質素穩定，不受季節影響；挑選成熟的草莓製成的酵母液，氣味香甜濃郁。

/葡萄乾酵母液/

材料

*葡萄乾......150克

水......300克

上白糖......23克

*要使用不含油脂的葡萄乾，含油的葡萄乾不會發酵，購買時請注意包裝上的標示。

培養酵母

玻璃瓶洗淨後，用沸水浸泡消毒，待玻璃瓶冷卻後使用。將葡萄乾、水和上白糖放進玻璃瓶中，蓋緊瓶子，放置在和暖陰涼處，避免陽光直射。每天搖晃瓶子2次，搖晃後打開瓶蓋，讓酵母液內裏的氣體排出轉入新鮮空氣。

發酵狀態

第二天葡萄乾吸收水分脹大，有些會浮起。第三至四天表面有很多小氣泡，而且葡萄乾會完全浮起，搖晃後打開瓶蓋，會聽到氣泡發出的「吱吱」聲，氣泡增多並且向上冒升，散發淡淡酒精氣味。

（接近完成了）第五至六天玻璃瓶靜止時表面有小氣泡，聽到氣泡發出的「吱吱」聲，瓶底會有白色的酵母菌沉澱物，搖晃後打開瓶蓋，氣泡增多並且向上冒升，散發強烈的酒精氣味。

葡萄乾酵母液完成了。

第一日

第二日

第四日

/草莓酵母液/

草莓……120克
水……300克
上白糖……23克

培養酵母

玻璃瓶洗淨後，用沸水浸泡消毒，待玻璃瓶冷卻後使用。將草莓、水和上白糖放進玻璃瓶中，蓋緊瓶子，放置在和暖陰涼處，避免陽光直射。每天搖晃瓶子2次，搖晃後打開瓶蓋，讓酵母液內裏的氣體排出轉入新鮮空氣。

發酵狀態

第二至三天水變成粉紅色，搖晃後打開瓶蓋，會聽到氣泡發出的「吱吱」聲，氣泡增多並且向上冒升，並散發出淡淡的草莓香。

（接近完成了）第四至五天表面有很多氣泡，玻璃瓶靜止時表面有氣泡，瓶底看到白色的酵母菌沉澱物，而且會聽到氣泡發出的「吱吱」聲，草莓顏色變白並浮起，搖晃後打開瓶蓋，氣泡增多並且向上冒升，這時你會嗅到濃郁的草莓香和輕微酒精的氣味。

草莓酵母液完成了。

第三日

第三日 玻璃瓶靜止時

第四日

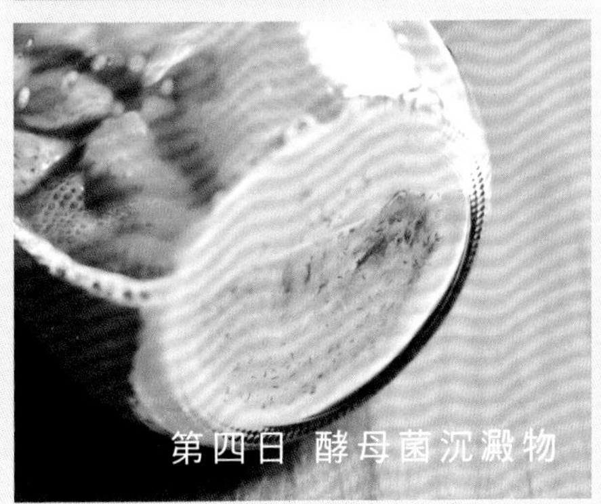
第四日 酵母菌沉澱物

完成了

保存酵母液

準備一個用沸水浸泡消毒的玻璃瓶，以濾網過濾出酵母液，放進瓶子，蓋緊瓶子放在雪櫃保存。可保存 3 個月，每個月取出，加入 1 小匙糖餵養即可。

把酵母液加入麵粉製作成狀態穩定充滿活力的原種，自家培育的天然酵母都是獨一無二，製作時可感受到天然酵母的生命力。

酵母菌會因應不同的環境、溫度、濕度等等的因素而有變化，配方沒有絕對的，可按酵母菌的活躍狀態和個人使用喜好把水和麵粉的用量比例作出調整，一步一步嘗試培養出健康活潑的天然酵母。

/原種（水量70%）/

第一天

酵母液50克、全麥粉50克放進容器中混合均勻，放置在室溫（約28℃）發酵約4小時至膨脹約兩倍，放進雪櫃保存。

第二天

從雪櫃取出容器，加入水35克、全麥粉50克，混合均勻，放置在室溫發酵至膨脹約兩倍，放進雪櫃保存。

第三天

從雪櫃取出容器，加入水35克、全麥粉50克，混合均勻，放置在室溫發酵至膨脹約兩倍，放進雪櫃冷藏一晚便可使用。

保存和續養

原種放進雪櫃保存，每次使用後要續養，取出多少便要補回多少，即使沒有使用，一星期也要續種1次以補給營養。

假設取出60克原種，便要補回水18克、高筋麵粉42克，混合均勻後，放置在室溫發酵至膨脹約兩倍，然後放進雪櫃保存。

計算方法：60 x 0.7=42克高筋麵粉　　60 - 42=18克水

持續續養時，高筋麵粉和全麥粉隔次使用；若感覺發酵力減弱，可在續養時把水轉換成酵母液。

乘0.7的原因是原種是70%水量

/烘焙百分比/

烘焙百分比是把配方中的麵粉總重量為100%，其他材料相當於麵粉重量的百分比。使用百分比，可把材料比例一目了然，如製作份量有改變時，按照百分比計算便可。

百分比計算方式範例

以下的範例是原來的麵粉總重量是200克，改變份量為麵粉總重量250克計算。

高筋麵粉	100%	200克
蜂蜜	1%	2克
上白糖	3%	6克
鹽	2%	4克
薯泥	25%	50克
水	35%	70克
橄欖油	5%	10克
原種	20%	40克

麵粉總重量為250克計算

高筋麵粉	100%	250克	
蜂蜜	1%	3克	< 250 x 1%=2.5克
上白糖	3%	8克	< 250 x 3%=7.5克
鹽	2%	5克	< 250 x 2%=5克
薯泥	25%	63克	< 250 x 25%=62.5克
水	35%	88克	< 250 x 35%=87.5克
橄欖油	5%	13克	< 250 x 5%=12.5克
原種	20%	50克	< 250 x 20%=50克

＊點數位以四捨五入計算

BUTTER ROLL

/小餐包/樸實無華，圓圓挺挺，可愛

可做12個小餐包

麵糰

高筋麵粉……300克　100%
上白糖……30克　10%
鹽……4.5克　1.5%
全蛋……37.5克　12.5%
水……135克　45%
牛油……30克　10%
原種……40克　30%

造型用

全蛋……適量

製作麵糰

牛油放在室溫至16-18℃。

原種先放在水中浸泡2-3分鐘，以打蛋器攪拌均勻後放進攪拌機之大盆內，放入所有材料（牛油除外），以慢速打至混合，再以高速打至麵糰表面光滑。加入牛油，以中速打至混合及柔滑狀態。

攪拌完成的溫度為23-24℃

用手把麵糰檢查，延展至薄如透明的薄膜即可，把麵糰塑成圓球狀，放入膠盆中。

第一次發酵

整盆麵糰放在28℃室溫中進行第一次發酵，至膨脹率達1.5倍大，需時約2小時。

視乎情況作出調整

翻麵

將麵糰由膠盆倒出，把麵糰四邊往中央摺，再放進膠盆中。

翻麵可使麵糰交換空氣，增加體積，並使表面更有彈性和緊致。

低溫發酵

整盆麵糰放進4-6℃雪櫃中，低溫發酵12-24小時至膨脹率達2.5倍。

回溫

從雪櫃取出整盆麵糰，放在28℃室溫中回溫0.5-1小時，要視乎情況作出調整。如從雪櫃取出時未完成發酵，可放在室溫至膨脹率達2.5倍為止。

分割・鬆弛

用刮板把麵糰準確分割成8份。用手輕壓，排出麵糰空氣，滾圓，蓋上濕布或保鮮盒，讓麵糰鬆弛20-30分鐘。

滾圓是把麵糰四邊往中央摺，摺口朝下，用手包着麵糰轉動，滾成表面緊致的圓球狀。

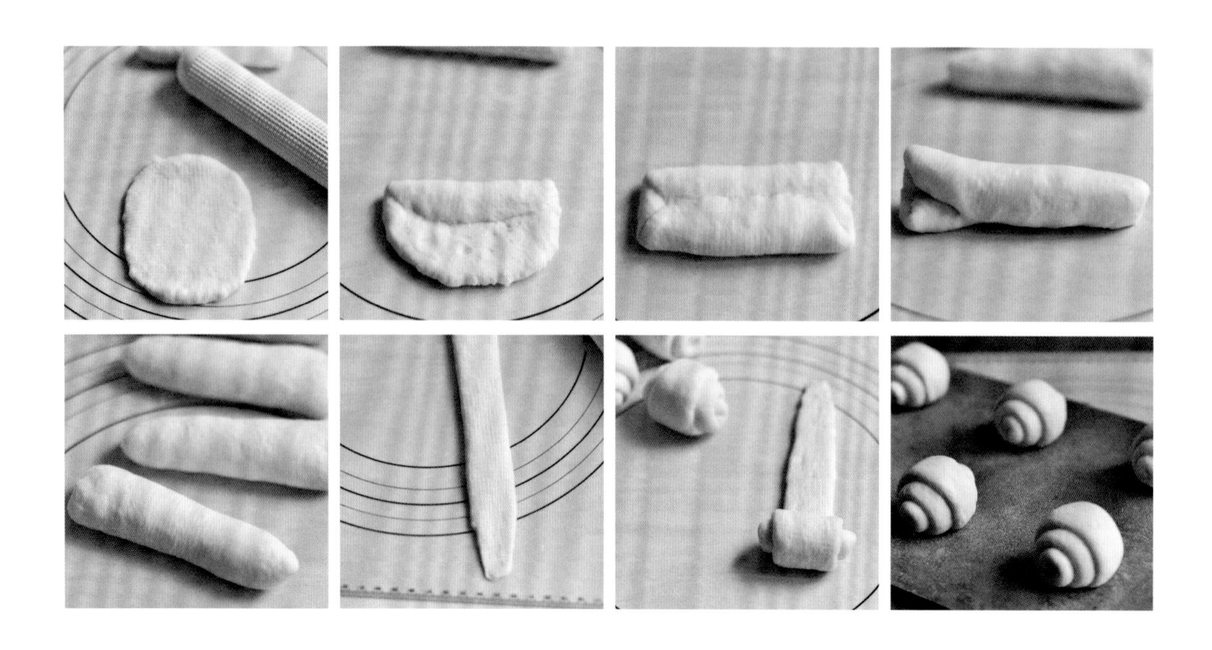

造型

用棍輕壓，排出麵糰空氣。把麵糰輕壓平，將上半部和下半部的麵糰往中央折入，再往下將收口處捏緊，將麵糰滾成12cm長的長圓柱形。

從麵糰較寬的一邊開始用木棍碾開至22-24cm長，碾開時一邊用手將麵糰往下拉。從較寬的一邊開始輕輕捲起，最後將收口處捏緊。

烤盤鋪上一張烘焙紙或不黏布，也可選用不黏的烤盤。麵糰收口處朝下，排放在烤盤上，因麵糰會膨脹，排放時要預留空間。

噴上水氣。

麵糰不宜含水量太高，造型時不宜捲得太緊，否則就會烤不出圓圓挺挺樣子可愛的小餐包。即使麵糰含水量不高，但添加了蛋有乳化作用，加上只要發酵得宜，小餐包口感也是十分軟綿。

第二次發酵

整盤麵糰放在28℃的溫度下約1小時，至膨脹率達2倍即完成最後發酵。完成最後發酵後，在麵糰表面用掃掃上蛋液。

蛋液是用全蛋打散製成的

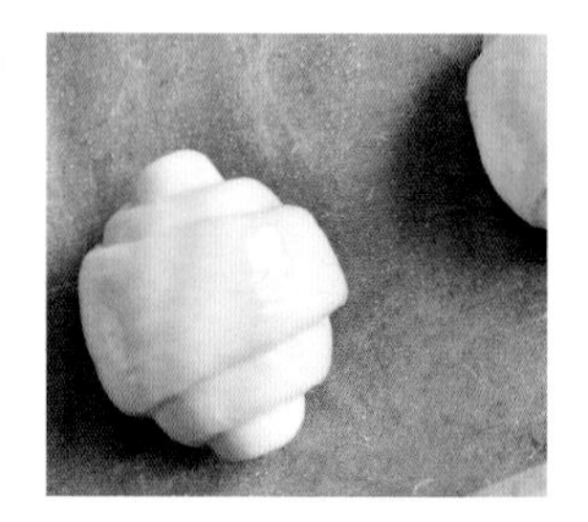

烘烤

整盤麵糰放進已預熱230℃的焗爐，再把焗爐溫度設定為200℃，烘烤10-15分鐘，烘烤完成後，立即把烤盤取出，把麵包放在架上待涼。

將小餐包切開中央，放進青瓜片和煙燻三文魚，另一個小餐包放進炒滑蛋、烤蘑菇和煙燻火腿，配上南瓜甘筍忌廉湯和烤南瓜沙律，在家也可簡單輕鬆做出豐富的早午餐。

SOYMILK ENGLISH MUFFIN

/豆乳馬芬/醇美，蘊含豆乳清香

可做7個直徑8cmx高2.5cm的豆乳馬芬

麵糰

高筋麵粉......70克 85%
米粉......30克 15%
上白糖.....8克 4%
鹽......3克 1.5%
豆乳......120克 60%
牛油......6克 3%
原種......60克 30%

造型用

粗粒玉米粉......適量

製作麵糰

牛油放在室溫至16-18℃。

原種先放在豆乳中浸泡2-3分鐘，以打蛋器攪拌均勻後放進攪拌機之大盆內，放入所有材料（牛油除外），以慢速打至混合，再以高速打至麵糰表面光滑。加入牛油，以中速打至混合及柔滑狀態。

攪拌完成的溫度為23-24℃

用手把麵糰檢查，延展至薄如透明的薄膜即可，把麵糰塑成圓球狀，放入膠盆中。

分割・鬆弛

用刮板把麵糰準確分割成8份。

用手輕壓，排出麵糰空氣，滾圓，蓋上濕布或放進保鮮盒，讓麵糰鬆弛20-30分鐘。

第一次發酵

整盆麵糰放在28℃室溫中進行第一次發酵，至膨脹率達1.5倍大，需時約2小時。

視乎情況作出調整

翻麵

將麵糰由膠盆倒出，把麵糰四邊往中央摺，再放進膠盆中。

翻麵可使麵糰交換空氣，增加體積，並使表面更有彈性和緊致。

低溫發酵

整盆麵糰放進4-6℃雪櫃中，低溫發酵12-24小時至膨脹率達2.5倍。

回溫

從雪櫃取出整盆麵糰，放在28℃室溫中回溫0.5-1小時，視乎情況作出調整。

如從雪櫃取出時未完成發酵，可放在室溫至膨脹率達2.5倍為止。

造型

用手輕壓麵糰，排出麵糰內的空氣，再次滾圓，在麵糰表面噴水，沾上粗粒玉米粉，放進已掃牛油的圓形模具中輕壓，再放進已鋪上一張烘焙紙或不黏布的烤盤上。

噴上水氣。

在圓形模具中掃上室溫牛油，可防止麵糰烘烤後黏模。

第二次發酵

整盤麵糰放在28℃的溫度下約1小時，至膨脹率達2倍，即完成最後發酵。完成最後發酵後，在麵糰上蓋上一張烘焙紙或不黏布，再壓上烤盤。

烘烤

放進已預熱210℃的焗爐內，放進烤盤後把焗爐溫度設定為190℃，烘烤13-18分鐘，烘烤完成後立即把烤盤取出，把麵包放在架上待涼。

SANDWICH BREAD

/吐司/最基本、最淳樸的滋味

麵糰

高筋麵粉......280克　100%
牛奶......182克　65%
淡忌廉......28克　10%
上白糖......11克　4%
鹽......4克　1.5%
牛油......11克　4%
原種......84克　30%

製作麵糰

牛油放在室溫至16-18℃。

原種先放在牛奶中浸泡2-3分鐘，以打蛋器攪拌均勻後放進攪拌機之大盆內，放入所有材料（牛油除外），以慢速打至混合，再以高速打至麵糰表面光滑。加入牛油，以中速打至混合及柔滑狀態。

攪拌完成的溫度為23-24℃

用手把麵糰檢查，延展至薄如透明的薄膜即可，把麵糰塑成圓球狀，放入膠盆中。

第一次發酵

整盆麵糰放在28℃室溫中進行第一次發酵，至膨脹率達1.5倍大，需時約2小時。

視乎情況作出調整

翻麵

將麵糰由膠盆倒出，把麵糰四邊往中央摺，再放進膠盆中。

翻麵可使麵糰交換空氣，增加體積，並使表面更有彈性和緊致。

低溫發酵

整盆麵糰放進4-6℃雪櫃中，低溫發酵12-24小時至膨脹率達2.5倍。

回溫

從雪櫃取出整盆麵糰，放在28℃室溫中回溫0.5-1小時，視乎情況作出調整。

如從雪櫃取出時未完成發酵，可放在室溫至膨脹率達2.5倍為止。

分割・鬆弛

用刮板把麵糰準確分割成3份，用手輕壓，排出麵糰空氣，滾圓，蓋上濕布或保鮮盒，讓麵糰鬆弛約20-30分鐘。

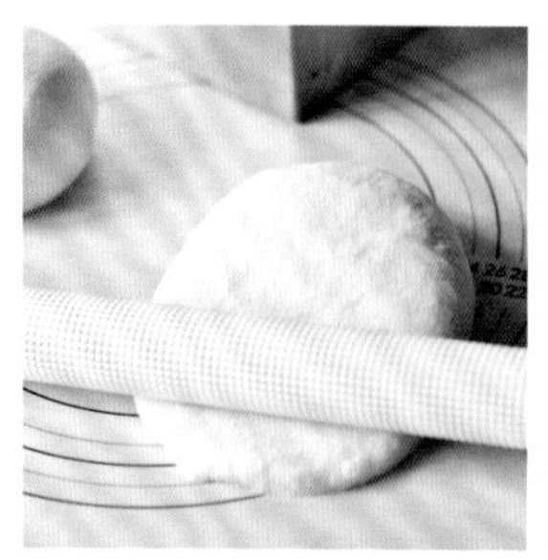

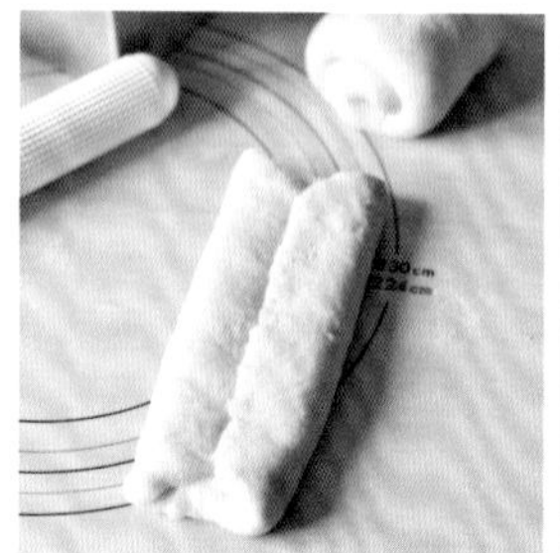

造型

用手輕壓麵糰，排出麵糰空氣，用棍將麵糰碾開成長方形，兩側往中央摺入，用手掌輕拍使麵糰緊貼。輕輕捲起，最後將收口處捏緊，把三個麵糰放進吐司模，噴上水氣。

第二次發酵

放在28℃的溫度下約1小時，至膨脹率達2倍即完成最後發酵。完成最後發酵後，在麵糰表面用掃掃上蛋液。

烘烤

放進已預熱230℃的焗爐，把焗爐溫度設定為200℃，烘烤30-35分鐘，烘烤完成後，立即把烤盤取出，把麵包放在架上放涼。

這是基本配方，做法和材料也十分簡單，完成品外皮香脆內裏濕潤柔軟。

成品在出爐後「霹哩啪啦」地唱歌，外皮慢慢爆出裂紋，形成輕脆口感。

吐司吃法百變，把吐司做成口袋三文治，配上菠菜、番茄片、金平牛蒡和多士，是方便快捷的早餐。

HAMBUGER Bun

/漢堡包/營養豐富的美式午餐，又帶點日式風味

可做8個直徑8cm x 高2.5cm的漢堡包

麵糰

高筋麵粉......160克　80%
法國麵包專用粉......40克　20%
上白糖......20克　10%
鹽......3克　1.5%
全蛋......20克　10%
水......128克　64%
牛油......16克　8%
原種......60克　30%

造型用

白芝麻......適量
全蛋......適量

製作麵糰

牛油放在室溫至16-18℃。

原種先放在水中浸泡2-3分鐘，以打蛋器攪拌均勻後放進攪拌機之大盆內，放入所有材料（牛油除外），以慢速打至混合，再以高速打至麵糰表面光滑。加入牛油，以中速打至混合及柔滑狀態。

攪拌完成的溫度為23-24℃

用手把麵糰檢查，延展至薄如透明的薄膜即可，把麵糰塑成圓球狀，放入膠盆中。

第一次發酵

整盆麵糰放在28℃室溫中進行第一次發酵，至膨脹率達1.5倍大，需時約2小時。

視乎情況作出調整

翻麵

將麵糰由膠盆倒出，把麵糰四邊往中央摺，再放進膠盆中。

翻麵可使麵糰交換空氣，增加體積，並使表面更有彈性和緊致。

低溫發酵

整盆麵糰放進4-6℃雪櫃中，低溫發酵12-24小時至膨脹率達2.5倍。

回溫

從雪櫃取出整盆麵糰，放在28℃室溫中回溫0.5-1小時，視乎情況作出調整。

如從雪櫃取出時未完成發酵，可放在室溫至膨脹率達2.5倍為止。

分割・鬆弛

用刮板把麵糰準確分割成8份，用手輕壓，排出麵糰空氣，滾圓，蓋上濕布或保鮮盒，讓麵糰鬆弛20-30分鐘。

造型

用手輕壓麵糰，把空氣排出，再次滾圓。放進已掃牛油的圓形模具中，輕壓，噴上水氣。

在圓形模具中掃上室溫牛油，是防止麵糰烘烤後黏黐模具。

第二次發酵

麵糰放在28℃的溫度約1小時，至膨脹率達2倍即完成最後發酵。完成最後發酵後，在麵糰表面用掃掃上蛋液，撒上白芝麻。

蛋液是用全蛋打散製成的。

烘烤

將麵糰放進已預熱210℃的焗爐，放進烤盤後把焗爐溫度設定為190℃，烘烤12-15分鐘。烘烤完成後，立即把烤盤取出，把麵包放在架上放涼。

將漢堡包對半切開，放上生菜、番茄片和自家製和牛漢堡扒，淋上照燒汁，配上炸洋蔥圈和薯角，WOW，美味！

PROVENÇALE FOUGASSE

/普羅旺斯/淳樸的南法鄉村滋味

麵糰

全麥粉......50克　25%
水......50克　25%
法國麵包專用粉......150克　75%
原種......40克　20%
鹽......2克　1%
水......90克　45%
橄欖油......12克　6%

灑面用

乾香草碎......適量

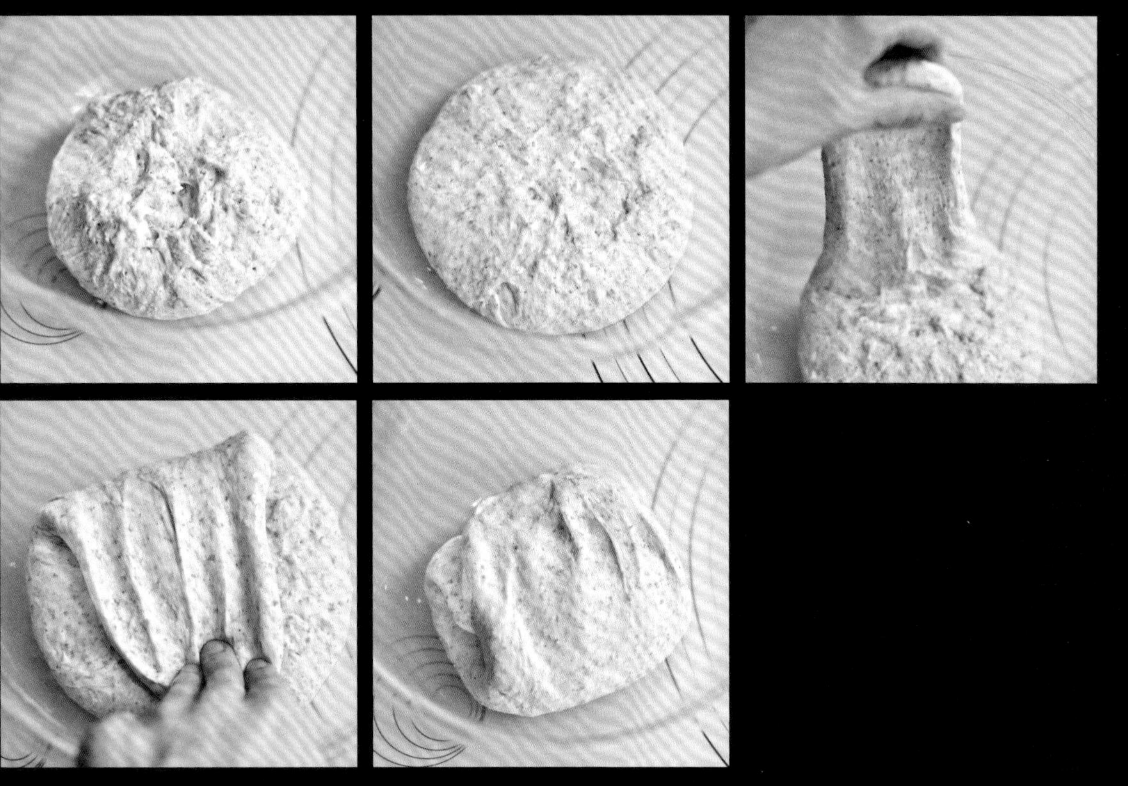

製作麵糰

全麥粉和50克水混合浸泡冷藏12-24小時。

原種放在90克水中浸泡2-3分鐘，攪拌均勻後加入全麥粉麵糰拌勻。加入法國麵包專用粉，用膠刮混合均勻，蓋上保鮮紙，放置在室溫20分鐘。

攪拌完成的溫度為20-22℃

加入鹽後，用手按壓麵糰至分佈均勻，雙手沾水，把麵糰周邊往中央摺入，蓋上保鮮紙，放置在室溫20分鐘。加入橄欖油，把麵糰周邊往中央摺入，重複動作直至混合均勻，蓋上保鮮紙，放置在室溫中30分鐘。

雙手沾水，把麵糰周邊往中央摺入，蓋上保鮮紙，放進4-6℃雪櫃中進行低溫發酵12-24小時，至膨脹率達2.5倍。

回溫

從雪櫃取出麵糰，放在28℃室溫中回溫0.5-1小時，視乎情況作出調整。

如從雪櫃取出時未完成發酵，可放在室溫至膨脹率達2.5倍。

造型

麵糰從膠盆倒出，拉成方型。用刮刀把麵糰分割成4份三角形。把麵糰對摺，用手指插進麵糰，使厚度成約0.6cm的三角形。麵糰放在不黏布上，倒入橄欖油，用滾刀切割花紋，輕輕把麵糰的切口拉開，灑上乾香草碎。

第二次發酵

將已造型的麵糰放在28℃的溫度下進行45分鐘最後發酵。

烘烤

放進已預熱250℃的焗爐，把焗爐溫度設定為220℃，烘烤8-12分鐘。烘烤完成後，立即把烤盤取出，把麵包放在架上放涼。

Fougasse是法國普羅旺斯的經典麵包，造型以「樹葉」最為多見，但也可以發揮自家創意，做出自己喜歡的形態。

簡單的添加香草，淳樸，不造作。Fougasse口感酥脆，充滿全麥麵粉的甘甜，配搭加入蒜頭和迷迭香烘烤的camembert cheese， 簡單但滋味無窮。

PESTO MONKEY BREAD

/羅勒青醬手撕包/嘗過一口，那清新的味道，味蕾都會歡呼

模具

長20cmx濶7.5cmx高6.5cm
長方模

麵糰

法國麵包專用粉......90克　50%
高筋麵粉......90克　50%
鹽......3克　1.5%
水......117克　65%
原種......36克　20%
橄欖油......5克　3%

羅勒青醬

新鮮羅勒葉......40克
已烘香的腰果......40克
巴馬臣芝士......20克
初榨橄欖油......40克
大蒜......1瓣

羅勒青醬做法

新鮮羅勒葉洗淨，瀝乾水分，不要讓水滴殘留在葉上。將所有材料放進攪拌器，攪拌成醬。將羅勒青醬放進乾淨已消毒的玻璃瓶內，放進雪櫃保存。

你可以做多一點羅勒青醬放在雪櫃保存，在玻璃瓶的面層倒入一層薄薄的橄欖油後再封存，可保存2星期及預防變色；也可做成冰粒存放在冰格，可保存數個月。

羅勒青醬不論是拌意粉或沾麵包也很好吃。材料比例可因個人口味作出改變，腰果可用松子代替。

製作麵糰

高筋麵粉和一半水混合浸泡冷藏12-24小時。

原種放在另一半水中浸泡2-3分鐘，攪拌均勻後加麵糰拌勻。加入法國麵包專用粉，用膠刮混合均勻，蓋上保鮮紙，放置在室溫20分鐘。

攪拌完成的溫度為20-22℃

加入鹽後，用手按壓麵糰至分佈均勻，雙手沾水，把麵糰周邊往中央摺入，蓋上保鮮紙，放置在室溫20分鐘。加入橄欖油，把麵糰周邊往中央摺入，重複動作直至混合均勻，蓋上保鮮紙，放置在室溫中30分鐘。

雙手沾水，把麵糰周邊往中央摺入，蓋上保鮮紙，放進4-6℃雪櫃中進行低溫發酵12-24小時，至膨脹率達2.5倍。

造型

麵糰從膠盆倒出，拉成方形，平均地塗上羅勒青醬，用手指插進麵糰使厚度成約1cm的方形，用刮刀把麵糰分割成5份，將麵糰重疊一起，按照模具的濶度切割成等份，把麵糰放進模具中，再用手輕輕整平。

第二次發酵

放在28℃的溫度下進行30分鐘最後發酵。

烘烤

放進已預熱230℃的焗爐，把焗爐溫度設定為200℃，烘烤25-30分鐘。

烘烤完成後，立即把烤盤取出，把麵包放在架上放涼。

撕一片麵包放上黑番茄、水牛芝士、羅勒青醬，磨些檸檬皮，加一點檸檬汁和橄欖油調味，很清新的味道。

FOCACCIA

FOCACCIA

/佛卡夏/迷迭香、黑橄欖，地中海風味躍然

佛卡夏（Focaccia）是義大利的經典扁形麵包。

它外皮輕脆，質地軟綿，組織佈滿大小不一的漂亮氣孔。

麵糰上可以依照自己的口味放上喜歡的材料，如蔬菜、芝士、香草、油浸番茄乾等等。多樣的變化，做出自家獨特的佛卡夏！

為了嘗到麵糰獨特的香氣，只用黑橄欖和迷迭香，簡單配搭，但隱隱啖出地中海風味。配方中添加馬鈴薯泥可提高麵糰的保濕效果，把麵糰揉至光滑時才下馬鈴薯泥和橄欖油，可使風味保持，不會散失。這個添加的時機，是決定麵包的香氣和味道的關鍵。

麵糰

高筋麵粉……200克　100%
蜂蜜……2克　1%
上白糖……6克　3%
鹽……4克　2%
薯泥……50克　25%
水……70克　35%
橄欖油……10克　5%
原種……40克　20%

灑面用

巴馬臣芝士……適量
黑橄欖……適量
橄欖油……適量
迷迭香……適量

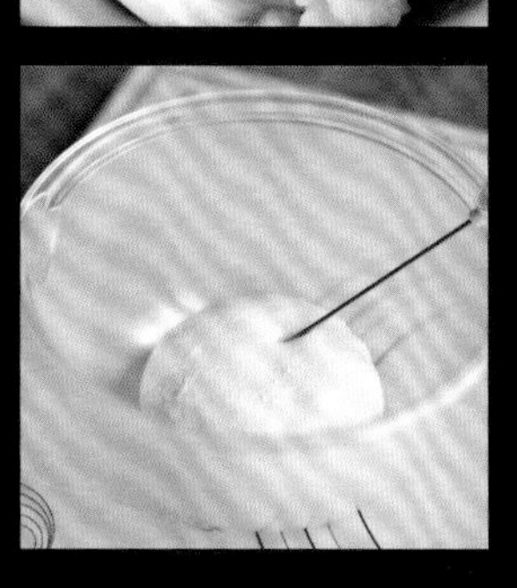

製作麵糰

薯仔洗淨，連皮蒸熟，取出，去皮，用叉壓成泥，待涼後放進雪櫃備用（使用前加入橄欖油混合）。

原種先放在水中浸泡2-3分鐘，以打蛋器攪拌均勻後放進攪拌機之大盆內，把所有材料（薯泥和橄欖油除外），以慢速打至混合，再以高速打至麵糰表面光滑。加入已混合的薯泥橄欖油，以中速打至混合及柔滑狀態。用手把麵糰檢查，延展至薄如透明的薄膜即可。把麵糰塑成圓球狀，放入膠盆中。

攪拌完成的溫度為23-24℃

這裏使用攪拌機製作麵糰，但也很適合手揉的方法，製作可參考P.33普羅旺斯（Fougasse）。

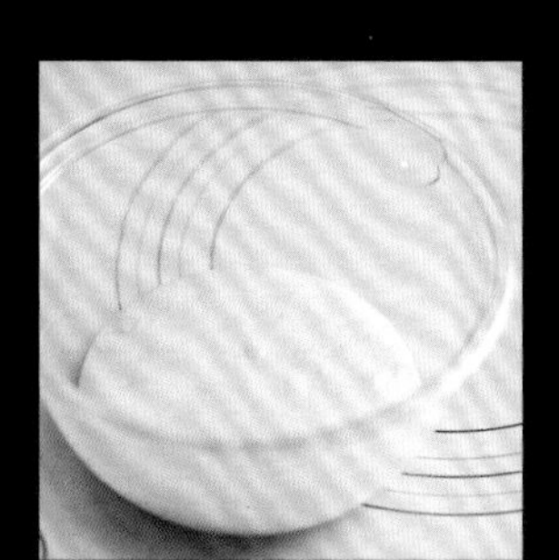

第一次發酵

放在28℃室溫中進行第一次發酵，至膨脹率達1.5倍大，約2小時。

視乎情況作出調整

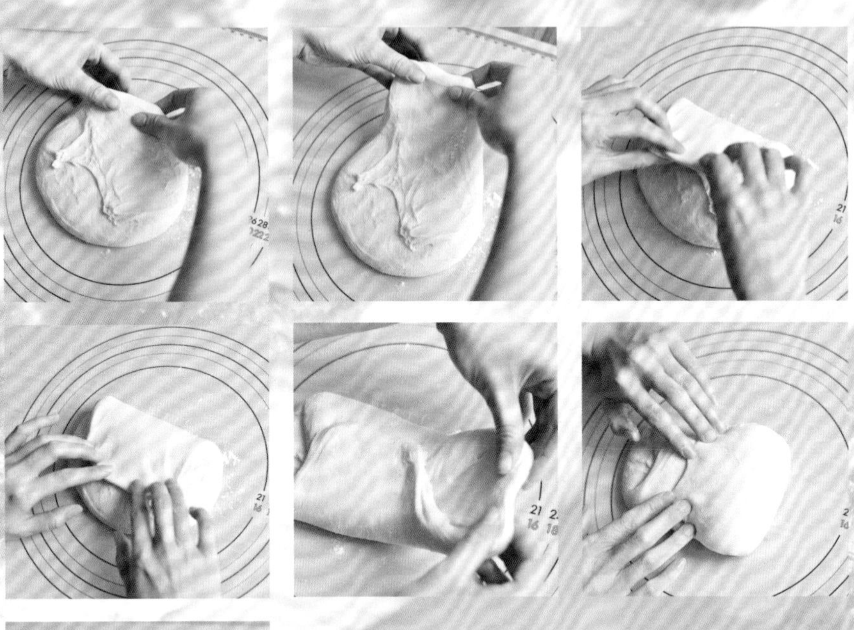

翻麵

將麵糰由膠盆倒出，把麵糰四邊往中央摺，再放進膠盆中。

翻麵可使麵糰交換空氣，增加體積，並使表面更有彈性和緊致。

低溫發酵

整盆麵糰放進4-6℃雪櫃中，低溫發酵12-24小時，至膨脹率達2.5倍。

回溫

從雪櫃取出整盆麵糰，放在28℃室溫中回溫0.5-1小時。

視乎情況作出調整

如從雪櫃取出時未完成發酵，可放在室溫至膨脹率達2.5倍為止。

造型

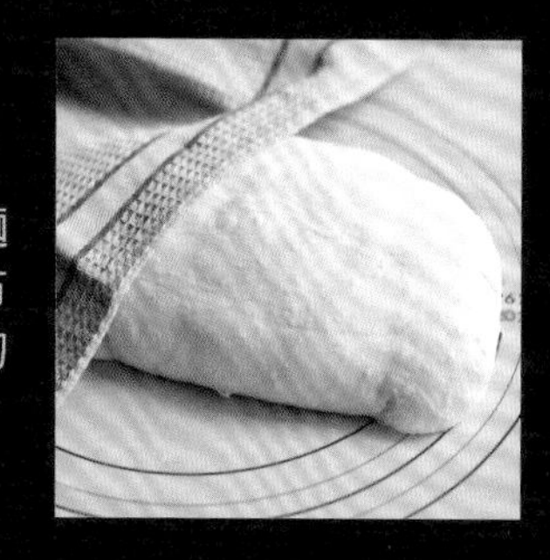

麵糰翻麵，蓋上濕布或保鮮盒，讓麵糰鬆弛20-30分鐘。麵糰放在不黏布上，倒入橄欖油，用手輕拍成厚度約1cm的長方形。

第二次發酵

放在28℃的溫度下進行30-40分鐘的最後發酵；用掃掃上橄欖油，用手指在麵糰表面壓出凹洞，放上迷迭香、黑橄欖和巴馬臣芝士。

烘烤

焗爐預熱280℃，把麵糰連同不黏布直接放在烤盤，把焗爐溫度設定為250℃烘烤15-20分鐘。烘烤完成後，把麵包放在架上待涼。

FLOUR TORTILLA

/墨西哥薄餅/ Tequila、墨西哥薄餅，拉丁的獨有風味

可做6個墨西哥薄餅

麵糰

高筋麵粉......75克　50%
全麥粉......75克　50%
鹽......1.5克　1%
水......53克　35%
酵母液......23克　15%

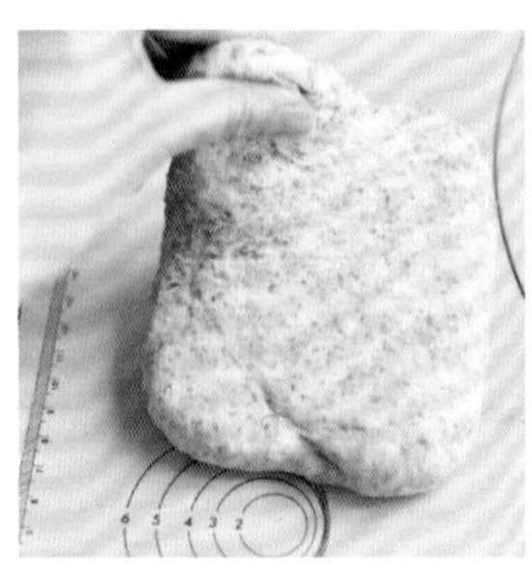

製作麵糰

全麥粉和水混合浸泡，冷藏12-24小時，然後把所有材料用手混合均勻，蓋上保鮮紙，放置在室溫中30分鐘。

把麵糰四邊往中央摺，再對摺，蓋上保鮮紙，放置在室溫中1小時。

也可將麵糰放進雪櫃冷藏，留待第二天才分割、造型。

分割・造型

用刮板把麵糰準確分割成6份，滾圓，讓麵糰鬆弛20-30分鐘，用棍碾成約20cm直徑的圓形。

燒成

鍋子以中火燒熱，放進一片麵糰，烘至稍微有膨脹及金黃色，反轉另一邊烘至上色。薄餅移出鍋子，用布蓋着保溫和保濕。

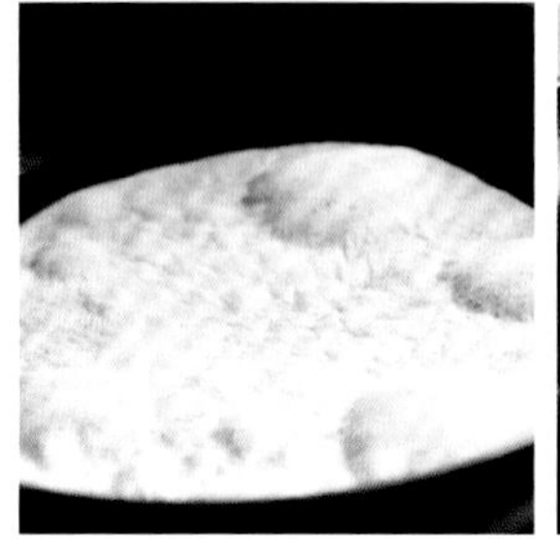

製作簡單方便，輕輕鬆鬆便可做出美味的墨西哥薄餅。配方中添加全麥麵粉，完成品散發獨特樸素麥氣，而且質地軟熟。

配上喜歡的材料如salsa，包捲起來，健康輕怡，Yummy！

PLAIN
BAGEL
/原味貝果/

CHOCOLATE
BAGLE
/朱古力貝果/

PLAIN BAGEL

/原味貝果/ 質感細致，耐嚼，味道醇厚香甜

可做6個原味貝果

麵糰

高筋麵粉......135克　45%
法國麵包專用粉......150克　50%
全麥粉......15克　5%
上白糖......15克　5%
鹽......4.5克　1.5%
水......135克　45%
原種......75克　25%

煮燙用

水......500克
蜂蜜......15克

製作麵糰

原種先放在水中浸泡2-3分鐘，以打蛋器攪拌均勻後，放進攪拌機之大盆內。把所有材料以慢速打至混合，再以中速打至麵糰表面光滑及柔滑狀態。用手把麵糰檢查，延展至薄如透明的薄膜即可。把麵糰塑成圓球狀，放入保鮮盒中，蓋緊蓋子。

攪拌完成的溫度為24-25℃

第一次發酵

麵糰放在28℃室溫中進行第一次發酵，至膨脹率達1.5倍大，約2小時。

視乎情況作出調整

低溫發酵

將麵糰放進4-6℃雪櫃中，低溫發酵12-24小時至膨脹率達2.5倍。

回溫

從雪櫃取出麵糰，放在28℃室溫中回溫0.5-1小時，視乎情況作出調整。

如從雪櫃取出時未完成發酵，可放在室溫至膨脹率達2.5倍為止。

分割・鬆弛

用刮板把麵糰準確分割成6份，用手輕壓，排出麵糰空氣，滾圓，蓋上濕布或保鮮盒，讓麵糰鬆弛約20-30分鐘。

造型

用手輕壓麵糰，排出空氣，用棍將麵糰碾開成長方形。從上方開始往下捲，將收口處捏緊，用手掌將麵糰輕輕搓長。將其中一端長約1cm的部分用木棍碾平，把麵糰扭轉兩次，將另一端包入，將收口處捏緊，噴上水氣。

不用進行第二次發酵，直接進行煮燙步驟。

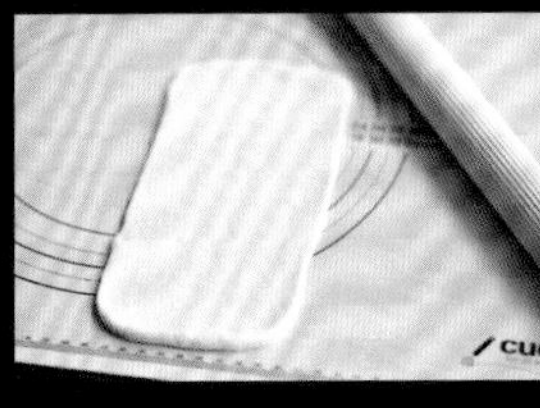
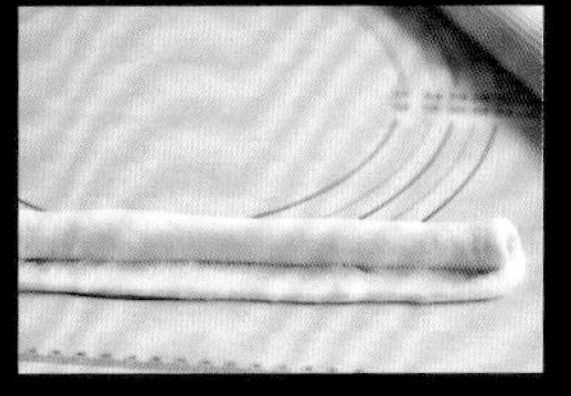

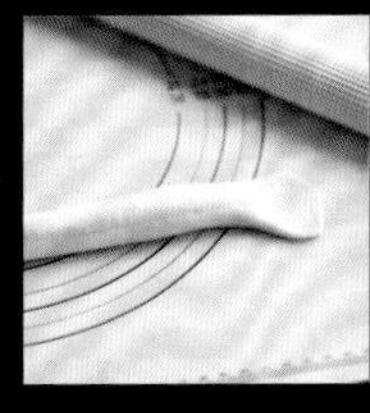

煮燙

鍋子放入水和蜂蜜，大火煮沸後，轉用中火將麵糰放入鍋子中，麵糰每邊煮燙20秒，用濾網撈起，把水瀝乾。麵糰放在烤盤上，收口處朝下。

因麵糰會膨脹，排放時要預留空間。

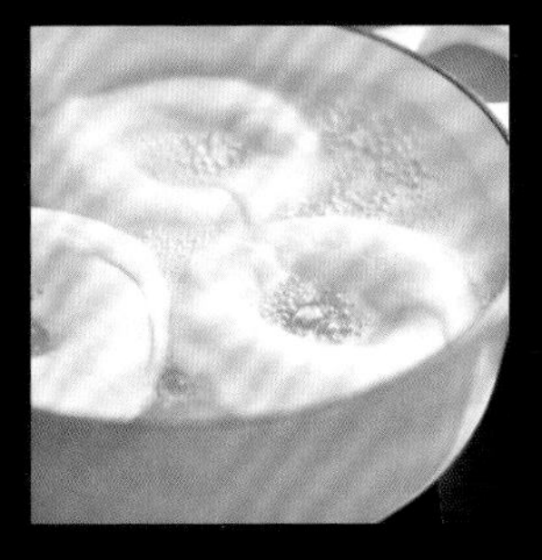

烘烤

烤盤放進已預熱230℃的焗爐內，立即把焗爐溫度設定為200℃，烘烤15-20分鐘。

貝果組織細緻緊密，口感耐嚼，長時間發酵使成品散發醇厚香甜的味道。

喜歡輕脆的外皮，成品在出爐後「霹哩啪啦」地唱歌，看着它的外皮爆出裂紋，很滿足。使用蜂蜜煮燙麵糰和高溫短時間烘焙有助產生裂紋。

將貝果對半切開，放上牛油生菜、碎蛋沙律和煙燻三文魚，自家製快速早餐，每天也可輕鬆做到。

CHOCOLATE BAGLE O

/朱古力貝果/啖啖濃郁的朱古力香，美味

使用基礎貝果麵糰，用手摺疊混入朱古力和可可粉，便可做成有漂亮紋路、味道更濃郁的朱古力貝果。

麵糰

高筋麵粉......150克 50%
法國麵包專用粉......150克 50%
上白糖......24克 8%
鹽......4.5克 1.5%
水......165克 55%
原種......75克 25%

混合用

65%朱古力......45克 15%
可可粉......15克 5%

煮燙用

水......500克
上白糖......15克

依照P.54「原味貝果」麵糰的製法

完成麵糰後，將已切碎的朱古力和可可粉鋪在麵糰上，用刮刀將麵糰切開再重疊，重複數次，直至混合大致均勻即可。

可是貝果麵糰含水量較低，混合期間如果感覺麵糰太乾，以致難以混合，可用噴水壺噴一點水在麵糰上才繼續混合。當然也可選擇直接把朱古力和可可粉加進攪拌機一起攪拌。

麵糰，就按照自己的喜好製作吧！

依照P.54「原味貝果」第一次發酵、低溫發酵、回溫、分割・鬆弛、造型、煮燙、烘烤的做法

FRENCH TOAST BAGEL

/法式貝果/朱古力貝果的變奏

法式吐司也不限定只採用吐司來做，甚麼種類的麵包也可！

可用已乾硬或放了幾天不新鮮的麵包來製作，因為新鮮或軟綿的麵包會吸收過多的蛋液，完成品口感會過於濕軟，如果真的是使用新鮮麵包，只要輕輕浸泡一下蛋液即可。

我用朱古力貝果做法式吐司。將貝果浸泡蛋液後放在膠盒冷藏保存，喜歡這樣的完成品不會過於濕軟！也可按照個人喜好，讓蛋液和麵包一同浸泡冷藏保存，做出口感濕潤的法式吐司！

我使用烘烤的方式製作，成品表面會有一點香脆；也可使用平底鑊煎至兩面金黃，成品軟綿。伴着香滑布包芝士，加上朱古力醬和莓醬汁同吃，成為美味的甜點。

材料

朱古力貝果......3個

蛋液

蛋......50克
糖霜......25克
蜂蜜......12克
牛奶......50克
忌廉......50克

醬汁

草莓......50克
紅桑子......50克
車厘子......50克
藍莓......50克
三溫糖......15克
砂糖......15克
檸檬汁......10克

朱古力醬

65%朱古力......20克
牛奶......8克

布包芝士

忌廉芝士......50克
砂糖......17克
檸檬汁......3克
無糖乳酪......50克
忌廉......50克

製作蛋液

將糖霜用濾網過篩。大碗中加入蛋、蜂蜜和糖霜，用打蛋器攪勻，逐少加入牛奶和忌廉，用打蛋器充分攪勻。

製作法式吐司

朱古力貝果每個切成6件，將朱古力貝果浸在蛋液中，底和面也充分吸收蛋液，放在保鮮盒中，放進雪櫃冷藏12-24小時。從雪櫃取出保鮮盒，將朱古力貝果排放在已鋪上烘焙紙或不黏布的烤盤上。放進已預熱230℃的焗爐，立即把焗爐溫度設定為200℃，烘烤10-15分鐘。烘烤完成後，立即取出烤盤，把麵包放在架上放涼。

製作醬汁

草莓切粒，車厘子、藍莓、紅桑子放進鍋子裏，加入三溫糖、砂糖和檸檬汁，以中火煮至軟身，熄火，備用。

製作朱古力醬

朱古力隔熱水至溶，牛奶隔熱水坐熱，把牛奶加入朱古力內，拌勻。

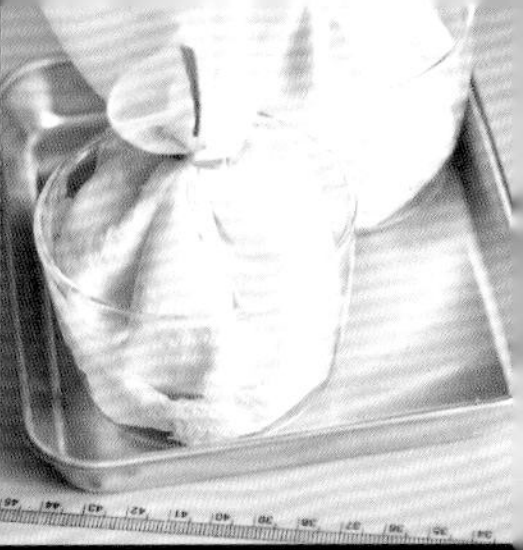

製作布包芝士

忌廉用電動打蛋器打起至6成；忌廉芝士回復室溫，將忌廉芝士放入大碗中，加入砂糖，用電動打蛋器打勻，分三次加入乳酪，用電動打蛋器打至混合。加入檸檬汁，用膠刮拌勻，最後加入忌廉，拌勻。

準備紗布；在器皿底部放一張抹手紙（大約摺成正方型），把紗布放進器皿中，倒入芝士糊，用橡筋或封口鐵絲綁緊，放在雪櫃冷藏至少12小時，食用時打開紗布，取出。

CROISSANT

/牛角/牛角，不能躲起享用，小心嘴角有碎屑

3*3*3牛角，可做13個

麵糰

法國麵包專用粉......200克　100%
上白糖......10克　5%
鹽......4克　2%
水......60克　20%
牛奶......50克　25%
原種......60克　30%
牛油......10克　5%

摺疊用牛油......130克　65%
蛋液......1個份量（塗面用）

製作麵糰

原種、牛奶和水用迷你打蛋器充分攪拌，加入其餘材料以慢速攪打至混合，再以中速打至麵糰表面大致光滑即可。

攪拌完成的溫度為20-21℃

將麵糰用保鮮紙包着，用木棍碾成約1cm厚正方形，放進冰箱-10℃冷藏3-4小時，取出放回雪櫃5℃冷藏至少8小時。

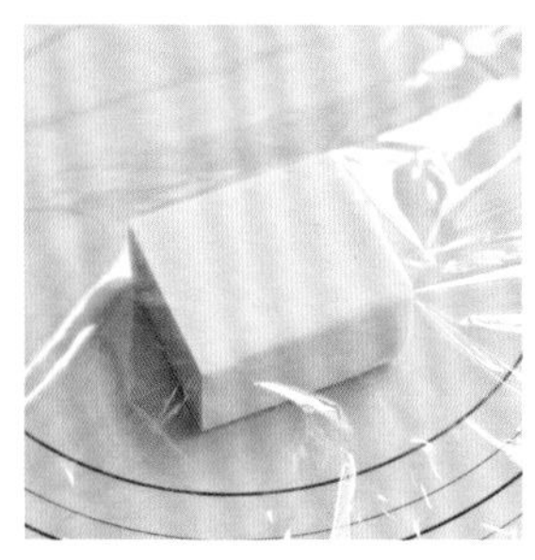

摺疊用牛油

牛油恢復室溫。

用保鮮紙包着，用棍碾成0.5cm厚正方形，放進雪櫃冷藏備用。

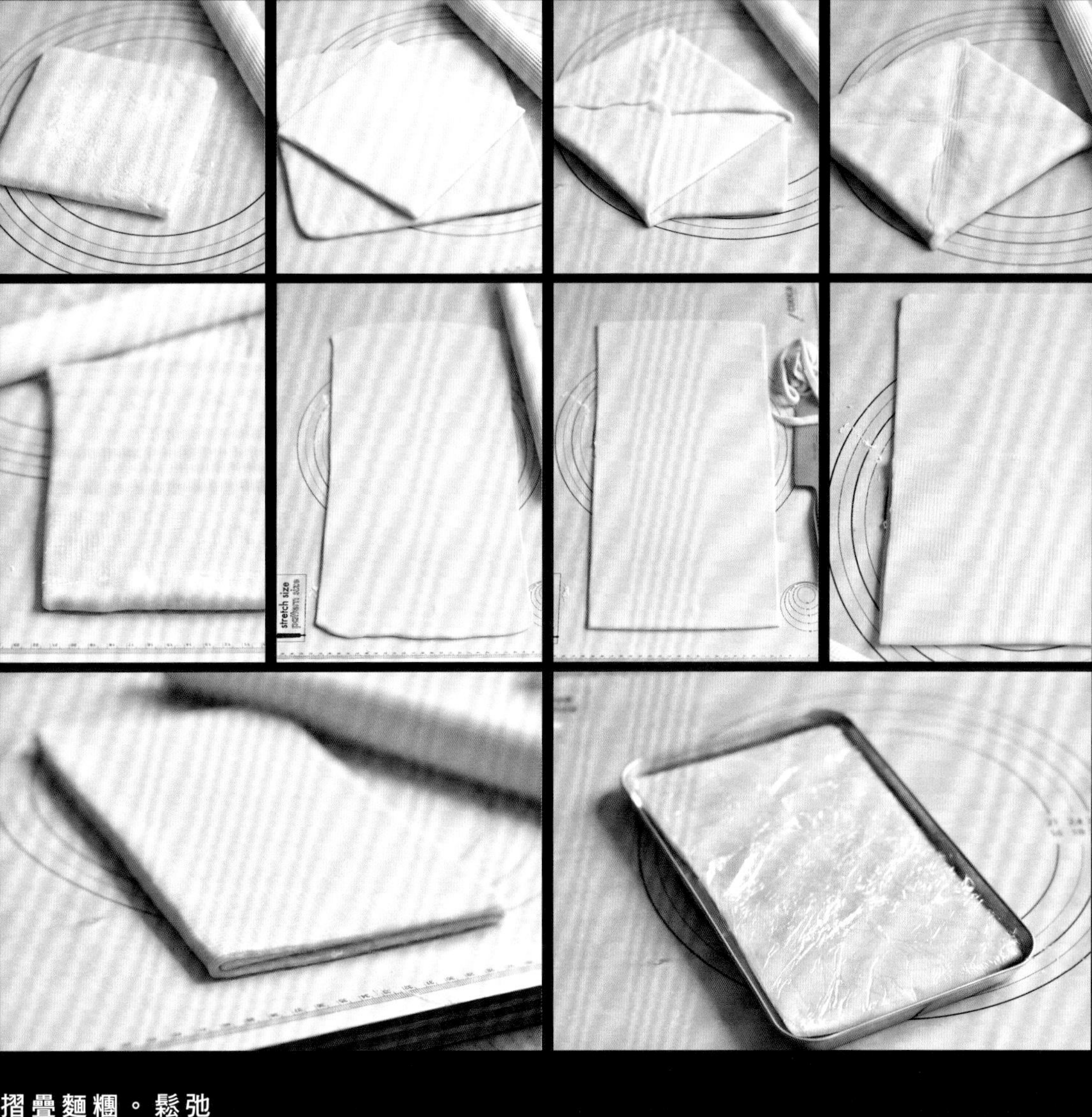

摺疊麵糰。鬆弛

從雪櫃取出摺疊用牛油，放置室溫至軟硬度跟麵糰相若。

用棍將麵糰碾開成厚度一致的正方形，放入摺疊用牛油，將麵糰的四個角摺起，麵糰輕微重疊，並且把接縫處黏和。

不要讓空氣混入，緊閉包入牛油片。

用棍將麵糰碾開成約0.3-0.5cm厚，用滾刀裁走四邊（約0.5cm邊位），摺成三摺，用木棍輕輕碾過使麵糰貼合。用保鮮紙將麵糰包好，冷藏鬆弛30分鐘至1小時，重複這步驟2次，共3次3摺。

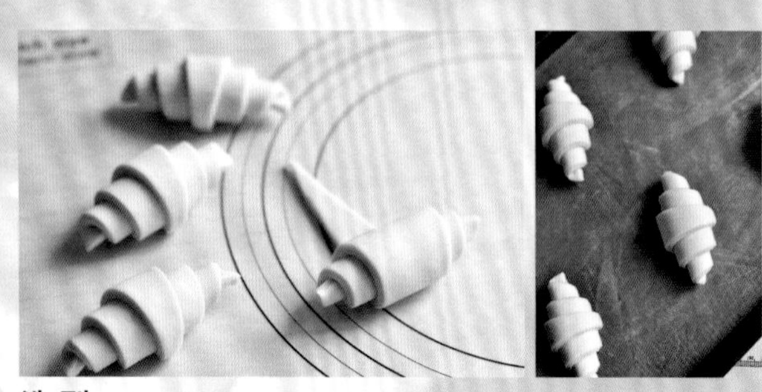

造型

從雪櫃中取出麵糰，用棍將麵糰碾開成約2.5mm厚，用滾刀裁走四邊（約0.5cm邊位），再用滾刀裁成底邊9cm*高度22cm的等腰三角形。在頂點抹一點水，從底邊開始輕輕捲起。

抹一點水可以幫助麵糰黏合

麵糰收口處朝下排放在烤盤上，噴上水氣。

麵糰會膨脹，排放時要預留空間。

烤盤鋪上一張烘焙紙或不黏布，也可選用不黏的烤盤。

第二次發酵

放在28℃的室溫約120分鐘至膨脹率達2倍即完成最後發酵，完成最後發酵後，在麵糰表面掃上蛋液。

蛋液是用全蛋打散製成的

烘烤

放進已預熱250℃的焗爐，放進烤盤後把焗爐溫度設定為220℃，烘烤12-13分鐘，烘烤完成後，立即把烤盤取出，把麵包放在架上待涼。

完成品的切面層次分明清晰，擁有迷人蜂巢組織，表面酥脆內裏濕潤鬆軟，
散發着香濃的牛油香，這就是令人著迷的牛角。
注意烘烤時必須烤至完全熟透，外皮烤至顏色略深，只要不是烤焦了就可。
如顏色烤得較淺的牛角，口感軟綿，不香也不酥脆。

BLACK SESAME CROISSANT

/黑芝麻牛角/芝麻加牛油香，你能抵抗誘惑嗎

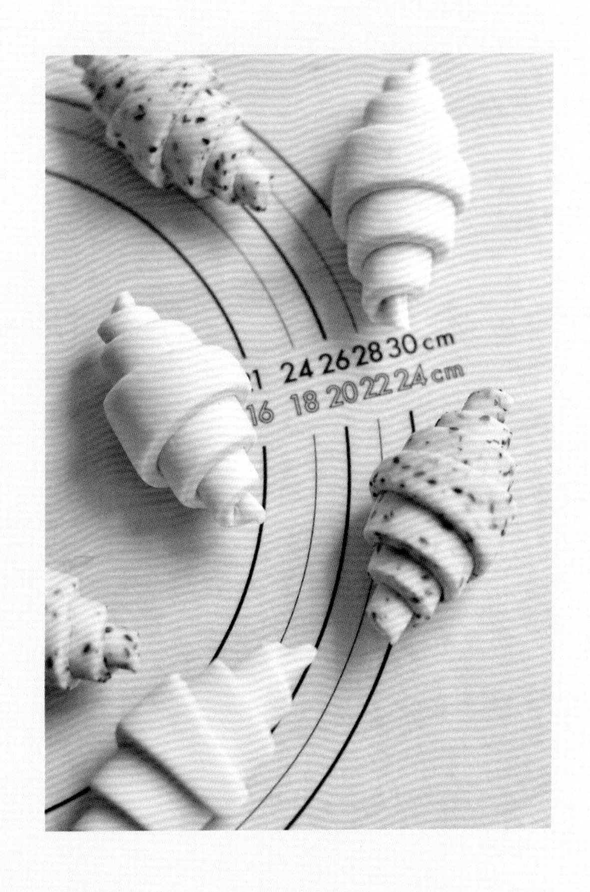

4*4牛角，可做13個

麵糰

法國麵包專用粉......200克　100%
黑芝麻......32克　16%
上白糖......10克　5%
鹽......4克　2%
牛奶......100克　50%
牛油......10克　5%
原種......60克　30%

摺疊用牛油......130克　65%
蛋液......1個份量（塗面用）

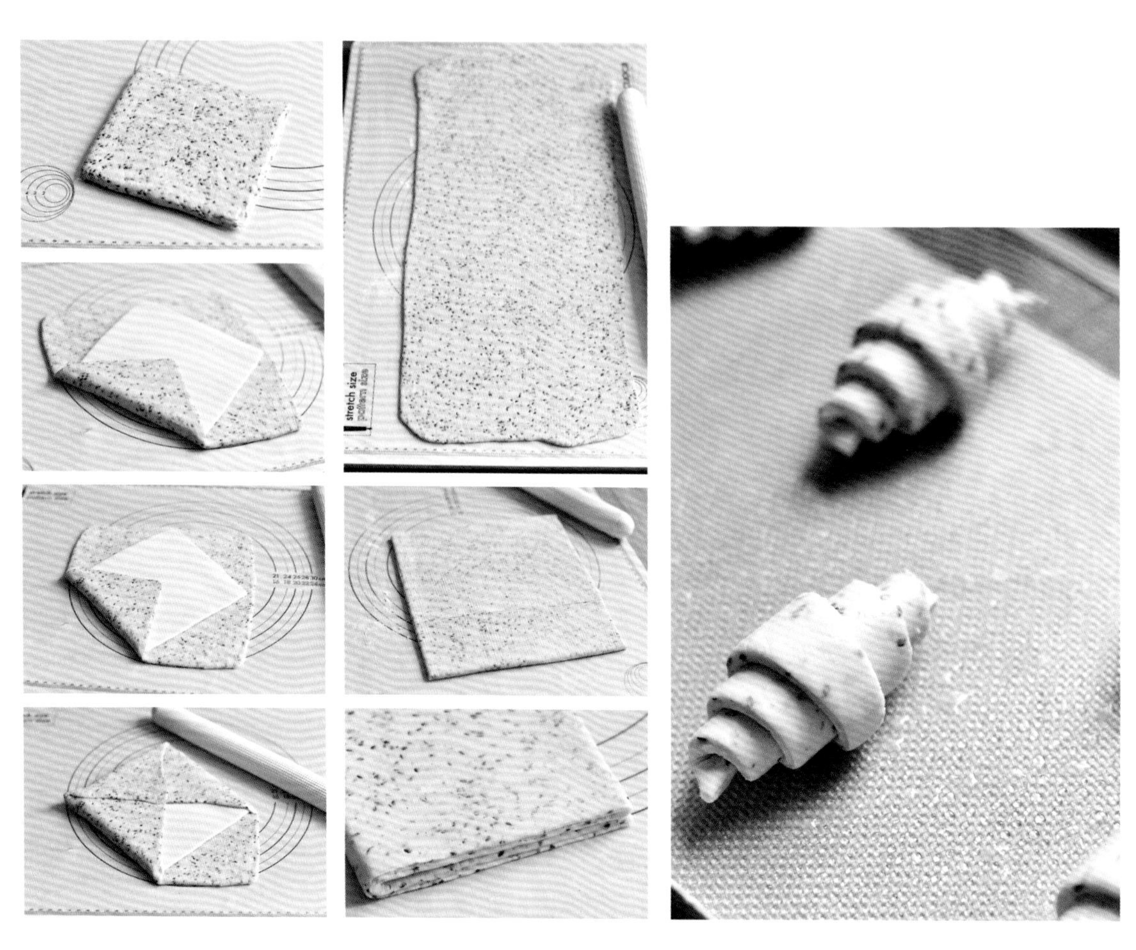

依照P.64「3*3*3牛角」麵糰的做法

摺疊麵糰。鬆弛

從雪櫃取出摺疊用牛油，放置室溫至軟硬度跟麵糰相若。

用棍將麵糰碾開成厚度一致的正方形，放入摺疊用牛油，將麵糰的四個角摺起，麵糰輕微重疊，並且把接縫處黏和。

不要讓空氣混入，緊閉包入摺疊用牛油。

用棍將麵糰碾開成約0.3-0.5cm厚，用滾刀裁走四邊（約0.5cm邊位），摺成四摺，用棍輕輕碾過使麵糰貼合。用保鮮紙將麵糰包好，冷藏鬆弛30分鐘至1小時，重複這步驟1次，共2次4摺。

造型

依照P.64「3*3*3牛角」的造型、第二次發酵和烘烤的做法

ALMOND CROISSANT

/杏仁牛角/ 把牛角變身成一道小甜點

牛角放上杏仁奶油餡，加上大量杏仁片烘烤後，伴着Tiramisu同吃，難以形容的美味。

牛角......4個

杏仁奶油

牛油......25克
糖霜......16克
杏仁粉......60克
全蛋......25克
低筋麵粉......5克

造型用

生杏仁片......適量

杏仁奶油

牛油和全蛋恢復室溫。不繡鋼盤中放入牛油和糖霜，用膠刮混合，加入粉類，再用膠刮混合；分數次加入全蛋，用打蛋器混合，冷藏待用。

造型

將牛角排放在烤盤上，平均地擠上杏仁奶油，放上生杏仁片。

烘烤

放進已預熱230℃的焗爐，把焗爐溫度設定為200℃，烘烤15-20分鐘。烘烤完成後，立即把烤盤取出，把麵包放在架上放涼。

STRAWBERRY DANISH

/草莓酥/ 酥名平實，但嘗一口，讓你難忘它的不凡滋味

依照P.64「3*3*3牛角」麵糰的材料和做法

吉士

牛奶……50克
蛋黃……10克
糖……10克
低筋麵粉……4克
雲呢拿條……1/6條
牛油……4克

製作吉士

牛油回復室溫。

用小刀剘雲呢拿條，刮籽；將蛋黃、糖和低筋麵粉放進大碗中用膠刮拌勻。

鍋子加入牛奶、雲呢拿籽和雲呢拿條，以中火煮至冒出蒸氣，逐少倒進蛋黃混合物內，用膠刮充分拌勻，用濾網過篩。再把混合物倒回煲中，以中火煮至濃稠。

期間不停用膠刮擦底攪拌，以防燒焦。

把吉士用濾網過篩，加入牛油，拌勻；用保鮮紙緊貼吉士的表面，放涼後，放進雪櫃冷藏1-2小時備用。

造型

從雪櫃中取出麵糰，再用木棍將麵糰碾開成約3mm厚，用滾刀裁走四邊（約0.5cm邊位），再用滾刀裁成10cm*10cm的正方形。在四角頂點抹一點水，把四角往中央摺入，每隻角稍微重疊按緊。

抹一點水可以幫助麵糰黏合

把麵糰放在8cm直徑的撻模，噴上水氣。

第二次發酵

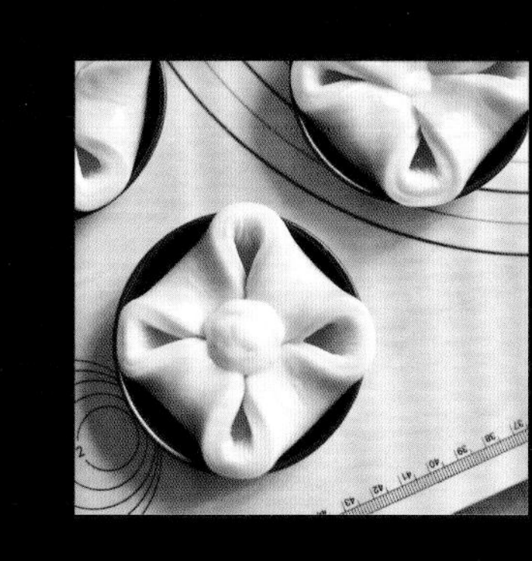

放在28℃的室溫約120分鐘至膨脹率達2倍即完成最後發酵，在麵糰表面掃上蛋液。

蛋液是用全蛋打散製成的

在中央擠上吉士。

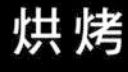

烘烤

放進已預熱250℃的焗爐，放進烤盤後把焗爐溫度設定為220℃，烘烤12-13分鐘，烘烤完成後，立即把烤盤取出，把麵包放在架上放涼。

放涼後，篩上糖霜，飾上草莓。

CINNAMON WALNUT BUNS

/肉桂核桃包/淡淡的核桃醬散發着肉桂香氣，夾雜咖啡糖和焦糖果仁，美味

可做8個肉桂核桃包

麵糰

高筋麵粉......180克　100%
上白糖......18克　10%
鹽......2克　1%
全蛋......90克　50%
牛油......54克　30%
原種......54克　30%

造型用

生杏仁片......適量

杏仁奶油

牛油......35克　19.4%
全蛋......35克　19.4%
糖霜......35克　19.4%
低筋麵粉......15克　8.3%
杏仁粉......35克　19.4%

肉桂核桃醬

核桃......120克　66.6%
牛油......20克　11%
肉桂粉......6克　3%
上白糖......35克　19.4%

咖啡糖霜

糖霜......50克
咖啡粉......6克
熱水......適量

其他材料

蛋液......1個份量（掃面用）
焦糖榛子
焦糖杏仁

杏仁奶油

牛油和全蛋回復室溫。不鏽鋼盤中放入牛油和糖霜，用膠刮混合，加入粉類，再用膠刮混合；分數次加入全蛋，用打蛋器混合，冷藏待用。

肉桂核桃醬

核桃放進已預熱170℃的焗爐，烘烤10-15分鐘，放涼待用。牛油恢復室溫；把所有材料放進食物處理器攪拌成醬，用保鮮紙包着碾成約0.5cm厚的片狀，冷藏待用。

製作麵糰

牛油放在室溫至16-18℃。

所有材料（牛油除外），以慢速打至混合，再以高速打至麵糰表面光滑。加入牛油，以中速打至混合及柔滑狀態。

攪拌完成的溫度為23-24℃

用手把麵糰檢查，延展至薄如透明的薄膜即可，把麵糰塑成圓球狀，放入膠盆中。

第一次發酵

整盆麵糰放在28℃室溫中進行第一次發酵，至膨脹率達1.5倍大，需時約2小時。

視乎情況作出調整

翻麵

將麵糰由膠盆倒出，把麵糰四邊往中央摺，再放進膠盆中。

翻麵可使麵糰交換空氣，增加體積，並使表面更有彈性和緊致。

低溫發酵

整盆麵糰放進4-6℃雪櫃中，低溫發酵12-24小時至膨脹率達2.5倍。

回溫

從雪櫃取出整盆麵糰，放在28℃室溫中回溫0.5-1小時。

如從雪櫃取時未完成發酵，可放在室溫至膨脹率達2.5倍為止。

分割。鬆弛

用手輕壓麵糰把空氣排出，分割出64克麵糰，再將麵糰分成8等分，滾圓後冷藏待用。

把餘下的麵糰拉成四角形，用棍將麵糰碾開成約3mm厚的長方形，用保鮮紙將麵糰包好，冷藏15-20分鐘。

造型

杏仁奶油回復室溫。

麵糰從雪櫃中取出，用抆刀平均地塗上杏仁奶油，左右和上端分別預留1cm空位，再放上厚片狀的肉桂核桃醬，將麵糰從上方往下捲起，並且把接縫處黏和，收口處朝下，切割成8等分。

將麵糰的切面朝上，把小麵糰從雪櫃中取出，按平至可覆蓋大麵糰的尺寸，黏在大麵糰的底部，排放在烤盤上，噴上水氣，用手掌壓平。

第二次發酵

放在28℃的室溫約60分鐘，完成最後發酵後，在麵糰表面掃上蛋液，放上生杏仁片。

烘烤

放進已預熱230℃的焗爐，把焗爐溫度設定為200℃，烘烤10-12分鐘。烘烤完成後，立即把烤盤取出，把麵包放在放涼架上放涼，擠上咖啡糖霜，放上焦糖榛子和杏仁。

製作咖啡糖霜

在麵糰烘烤時製作咖啡糖霜

把咖啡粉注入少量熱水，拌至溶解；糖霜用濾網過篩，逐少加入咖啡水，用膠刮拌勻，直至成濃稠狀即完成，在器皿上蓋上濕布，待用。

BRIOCHE

/布里歐吐司/抗拒不了的牛油香

發酵種

法國麵包專用粉......52克　20%
原種......7克　2.5%
牛奶......40克　15%

製作發酵種

原種放在牛奶中浸泡2-3分鐘，攪拌均勻後加入法國麵包專用粉拌勻。放置在28℃室溫中發酵3小時，再放進4-6℃雪櫃中冷藏發酵18-24小時。

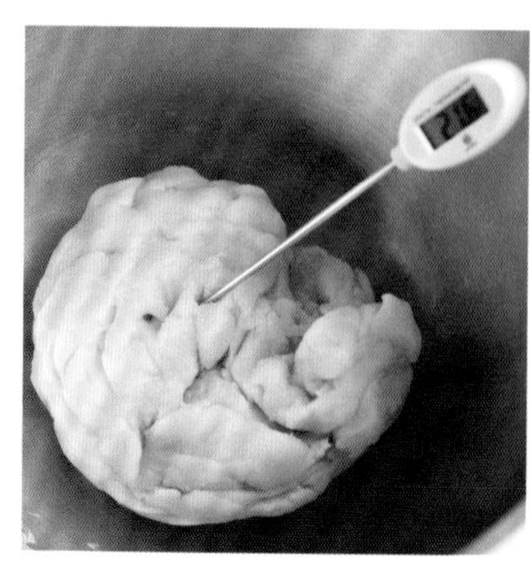

自我分解麵糰

法國麵包專用粉......78克　30%
高筋麵粉......130克　50%
上白糖......26克　10%
蛋黃......78克　30%
淡忌廉......52克　20%
牛油......52克　20%

製作自我分解麵糰

牛油回溫至16-18℃。

把所有材料放入攪拌機之大盆內，攪拌至混合均勻。

攪拌完成的溫度為21-22℃

把麵糰放進已放有烘焙紙的膠盒內，放進4-6℃雪櫃中冷藏18-24小時。

自我分解法讓麵糰充分水合形成筋度，製作麵糰時可減少攪拌時間，也可避免麵糰溫度過高，麵糰溫度過高會使牛油溶化，以致完成品較乾，失去應有的濕潤口感。

主麵糰

原種......72克　27.5%
鹽......5克　2%
牛油......104克　40%
牛奶......26克　10%

製作主麵糰

牛油回溫至16-18℃。

把發酵種和自我分解麵糰撕成小塊放入攪拌機之大盆內，加入原種、鹽和牛奶。以慢速打至混合，再以高速打至麵糰表面光滑。加入一半份量的牛油，以中速打至大致混合。再加入餘下的牛油，以中速打至混合及柔滑狀態。用手把麵糰檢查，延展至薄如透明的薄膜即可。

攪拌完成的溫度為22-23℃

把麵糰塑成圓球狀，放入膠盆中。

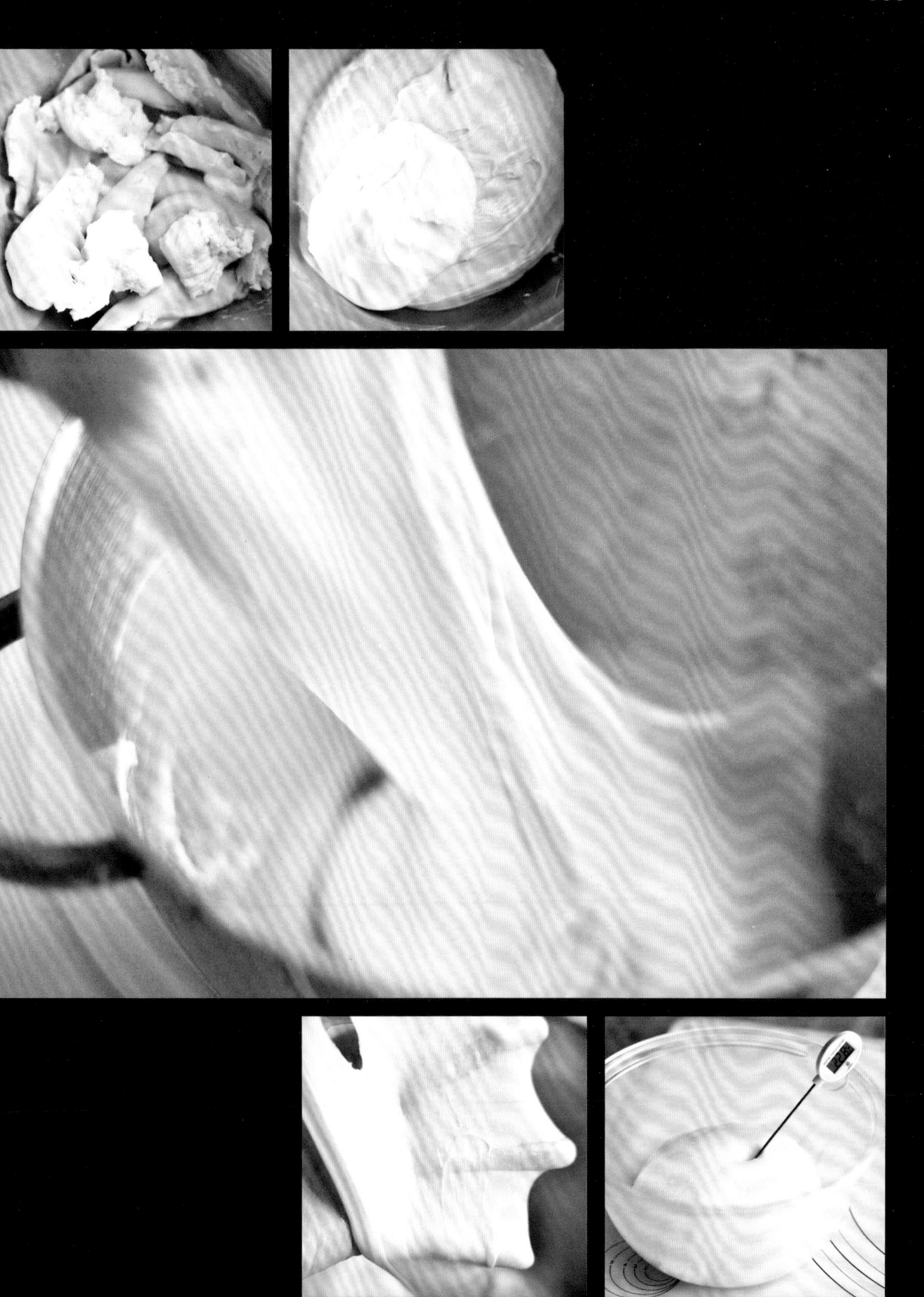

第一次發酵

放在28℃室溫中進行第一次發酵，至膨脹率達2倍大，約3-4小時。

視乎情況作出調整

排氣・冷卻

將麵糰按壓排氣，放在已鋪上烘焙紙的烤盤上，包上保鮮紙，放進-10℃的冰箱冷藏3小時以上（3天內使用）。

冷藏太久會使麵糰變硬，取出放回4-6℃雪櫃冷藏至少8小時，再進行製作。

分割・鬆弛

用刮板把麵糰準確分割成12份；用手輕壓麵糰，把空氣排出，滾圓，放在已鋪上烘焙紙的烤盤上，包上保鮮紙，放進4-6℃雪櫃中，冷藏鬆弛20-30分鐘。

造型

用手輕壓，把麵糰排出空氣，再次滾圓；6個麵糰一組，交錯放進模具中。

第二次發酵

放在28℃的室溫約120分鐘至膨脹率達2倍，即完成最後發酵。

完成最後發酵後，在麵糰表面掃上蛋液。

蛋液是用全蛋打散製成的

烘烤

放進已預熱220℃的焗爐，放進烤盤後把焗爐溫度設定為200℃，烘烤30-35分鐘。烘烤完成後，把麵包脫模，放在放涼架上放涼。

Brioche是比較像甜點的麵包，配方中含有大量蛋和牛油，成品組織像棉花般輕盈柔軟，口感潤澤，散發着牛油濃郁香氣。伴上藍莓果醬，更相得益彰。

COFFEE MONKEY BREAD

/咖啡紅莓手撕包/隨心、隨意享用，展開美好的一天

麵糰

高筋麵粉......200克　100%
三溫糖......20克　10%
鹽......2克　1%
全蛋......30克　15%
原種......60克　30%
牛奶......90克　45%
牛油......30克　15%
紅莓乾......40克　20%

咖啡糖漿

咖啡粉......10克　5%
熱水......15克　7.5%
牛油......25克　12.5%
三溫糖......25克　12.5%
杏仁......30克　15%

其他用具材料

方形餅模（6.5cm）......3個
黃金麵包模（直徑8.5cmx高6.5cm）......3個
粗粒玉米粉

製作咖啡糖漿

杏仁放進已預熱170℃的焗爐，烘烤8-10分鐘，待涼後切碎待用。

咖啡粉加入熱水中，拌勻，備用。

牛油隔熱水坐至溶，加入三溫糖拌勻，再注入咖啡水拌勻；使用前加入杏仁碎。

如在造型前製作咖啡糖漿，放置室溫太久牛油會凝固，可將它隔熱水加熱至溶解便可使用。

製作麵糰

牛油放在室溫至16-18℃。

原種先放在牛奶中浸泡2-3分鐘，以打蛋器攪拌均勻後放進攪拌機之大盆內，放入所有材料（牛油除外），以慢速打至混合，再以高速打至麵糰表面光滑。加入牛油，以中速打至混合及柔滑狀態。

攪拌完成的溫度為23-24℃

用手把麵糰檢查，延展至薄如透明的薄膜即可，把麵糰塑成圓球狀，放入膠盆中。

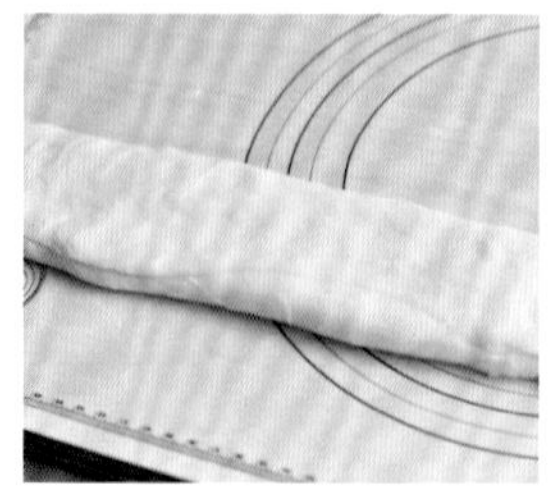
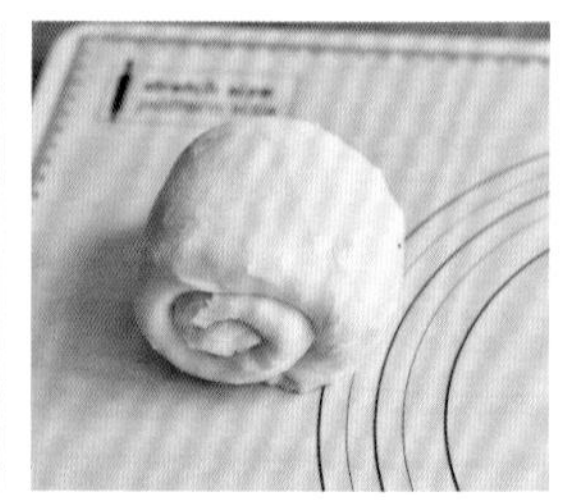

紅莓乾在使用前先浸泡在水30分鐘（配方外），再放在網篩瀝乾10分鐘。

將攪拌完成的麵糰拉成正方形放在工作板上，放上紅莓乾，捲起，用手壓平，再捲起，把麵糰塑成圓球狀，放入膠盆中。

整盆麵糰放在28℃室溫中進行第一次發酵，至膨脹率達1.5倍大，需時約2小時。

視乎情況作出調整

翻麵

將麵糰由膠盆倒出，把麵糰四邊往中央摺，再放進膠盆中。

翻麵可使麵糰交換空氣，增加體積，並使表面更有彈性和緊致。

低溫發酵

整盆麵糰放進4-6℃雪櫃中，低溫發酵12-24小時至膨脹率達2.5倍。

回溫

從雪櫃取出整盆麵糰，放在28℃室溫中回溫0.5-1小時，視乎情況作出調整。

如從雪櫃取時未完成發酵，可放在室溫至膨脹率達2.5倍為止。

分割

用刮板把麵糰隨意分割成小份，用手輕壓，排出麵糰空氣，滾圓。

造型

把麵糰放進咖啡糖漿中，取出；在麵糰表面噴水，沾上粗粒玉米粉，放進已掃牛油的模具中輕壓，噴上水氣。

在模具中掃上室溫牛油，可防止麵糰烘烤後黏模。

第二次發酵

放在28℃的室溫約60分鐘至膨脹率達2倍，即完成最後發酵。完成最後發酵後，在麵糰上蓋一張烘焙紙或不黏布，再放上烤盤。

烘烤

放進已預熱210℃的焗爐，把焗爐溫度設定為190℃，烘烤13-18分鐘，烘烤完成後，立即把麵包取出放在架上放涼。

簡單隨意的造型，完成品樣子可愛獨特，
散發着濃郁的咖啡香氣；
伴着dulce de leche吃，更是美味。

Dulce de leche

Dulce de leche（牛奶焦糖醬）是西班牙的傳統食物。

Dulce是焦糖，leche是牛奶的意思。

Dulce de leche用途廣泛，可以塗抹吐司、餅乾，做甜點或拌入朱古力、咖啡等等飲品中。

它的味道多變，可加入肉桂粉、咖啡粉、朗姆酒、朱古力等等，做出各式各樣的口味。

Dulce de leche可用煉奶或用牛奶加糖製作，做法簡單，製成品質地濃稠，十分幼滑，呈焦糖色，香氣濃郁。

以下是用煉奶製作Dulce de leche：

蒸爐做法

把一罐煉奶放進蒸爐，以100℃蒸3小時，完成後讓它待在爐中至完全冷卻，放入雪櫃冷藏，至少可以存放1個月。開封後，放進玻璃瓶，可以存放1星期。

蒸鍋做法

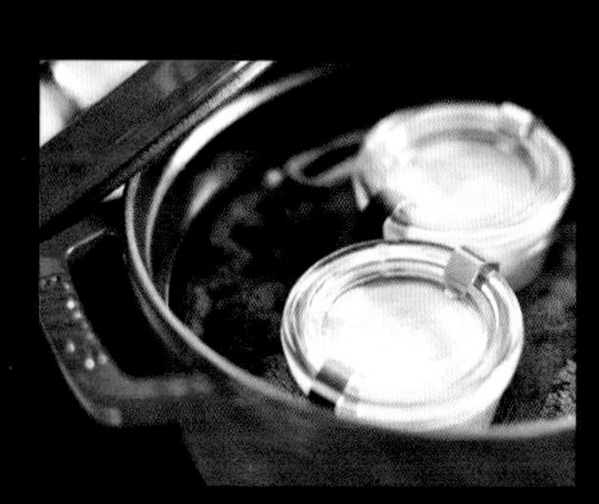

玻璃瓶洗淨後，用沸水浸泡消毒，待玻璃瓶冷卻後使用。把煉奶倒入玻璃瓶，蓋緊瓶子。

鍋裏放一條毛巾（防止玻璃瓶在烹煮過程中搖晃），放上玻璃瓶，注入瓶子三分一的水，蓋上鍋蓋。以中小火烹煮，保持微沸的狀態，烹煮約3-4小時。

完成後讓它待在鍋中至完全冷卻，放入雪櫃冷藏，至少可以存放1個月。

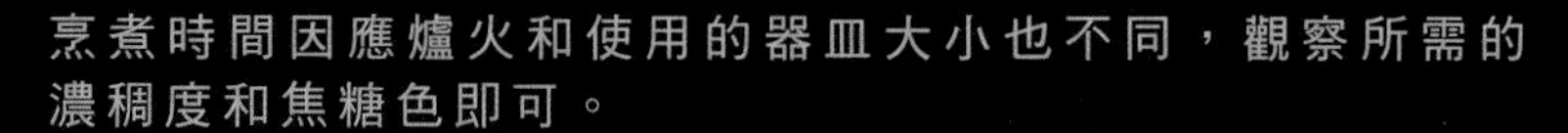

烹煮時間因應爐火和使用的器皿大小也不同，觀察所需的濃稠度和焦糖色即可。

焗爐做法

焗爐預熱200℃。

把煉奶倒入可進烤箱的器皿中（最好使用底部寬一點的器皿，可縮短烘烤時間）。用鋁箔紙封緊表面，放在烤盤中並注入熱水至器皿的一半高度。放進焗爐，以180℃烘烤至少1.5小時，每30分鐘取出用膠刮攪拌一次。

完成後趁熱放進玻璃瓶，待涼後放雪櫃冷藏，可保存1星期。

烘烤時間因應焗爐和使用的器皿大小也有不同，觀察所需的濃稠度和焦糖色即可。

BLACK SESAME LOAF

/黑芝麻吐司/芝麻香味饞人，只嘗一口，未能滿足

麵糰

高筋麵粉......140克　100%
三溫糖......17克　12%
鹽......2克　1.5%
全蛋......14克　10%
水......73克　52%
牛油......18克　13%
原種......42克　30%

黑芝麻夾心

無糖黑芝麻醬......70克　50%
牛奶......49克　35%
黑芝麻......30克　22%
糖......10克　7%
低筋麵粉......20克　14%
牛油......5克　3.5%

製作黑芝麻夾心

低筋麵粉用濾網過篩；牛油恢復室溫；無糖黑芝麻醬、糖及低筋麵粉拌勻，分數次加入牛奶，用膠刮拌勻。

每次加入牛奶前，要確實充分拌勻才再加入。

用微波爐500w 加熱以上的黑芝麻牛奶麵糰40秒，取出用膠刮拌勻成糕狀，再以500w 加熱40秒，取出，加入牛油拌勻。用保鮮紙包着再碾成15x15cm片狀，冷藏待用。

製作麵糰

牛油放在室溫至16-18℃。

原種先放在水中浸泡2-3分鐘，以打蛋器攪拌均勻後放進攪拌機之大盆內，放入所有材料（牛油除外），以慢速打至混合，再以高速打至麵糰表面光滑。加入牛油，以中速打至混合及柔滑狀態。

攪拌完成的溫度為23-24℃

用手把麵糰檢查，延展至薄如透明的薄膜即可，把麵糰塑成圓球狀，放入膠盆中。

第一次發酵

整盆麵糰放在28℃室溫中進行第一次發酵，至膨脹率達1.5倍大，需時約2小時。

視乎情況作出調整

翻麵

將麵糰由膠盆倒出，把麵糰四邊往中央摺，再放進膠盆中。

翻麵可使麵糰交換空氣，增加體積，並使表面更有彈性和緊致。

低溫發酵

整盆麵糰放進4-6℃雪櫃中，低溫發酵12-24小時至膨脹率達2.5倍。

回溫

從雪櫃取出整盆麵糰，放在28℃室溫中回溫0.5-1小時，視乎情況作出調整。

如從雪櫃取時未完成發酵，可放在室溫至膨脹率達2.5倍為止。

摺疊麵糰。鬆弛

用手輕壓麵糰排出空氣，把麵糰拉成四角形；用棍將麵糰碾開成厚度一致的正方形，放入黑芝麻夾心。將麵糰的四個角摺起，麵糰輕微重疊，並且把接縫處黏合。

用棍將麵糰碾開成約1cm厚，摺成四摺，用棍輕輕碾過使麵糰貼合。

重複這步驟一次

將麵糰冷藏，鬆弛20-30分鐘。

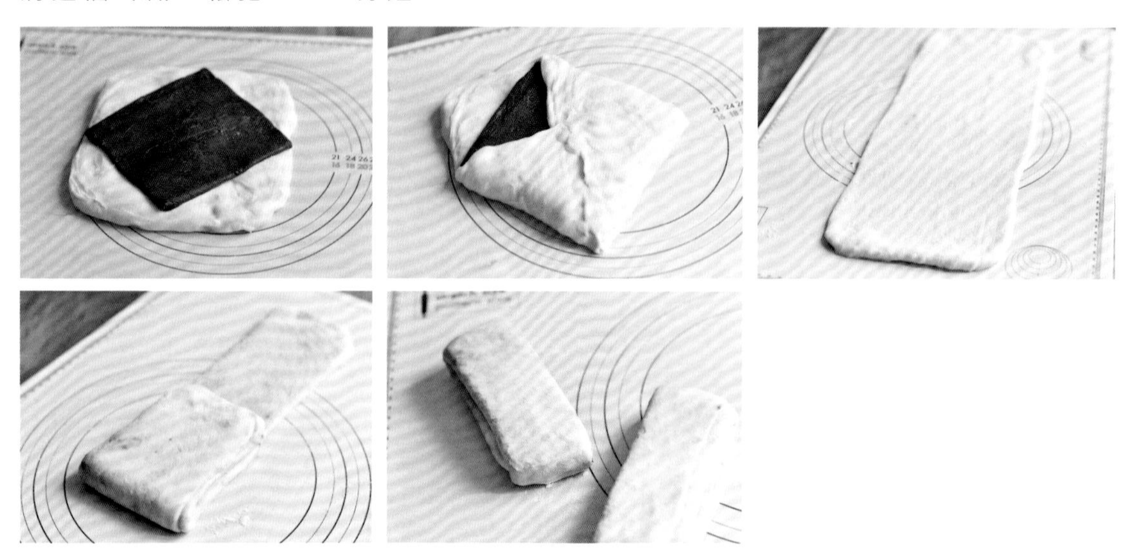

造型

從雪櫃中取出麵糰，用棍輕輕碾過使麵糰厚度一致，分割成6等分。每三條為一組編成三手辮，捲起，放進吐司模，噴上水氣。

第二次發酵

放在28℃的室溫約60分鐘至膨脹率達2倍即完成最後發酵，蓋上吐司模的蓋。

烘烤

放進已預熱230℃的焗爐，把焗爐溫度設定為200℃，烘烤25-30分鐘，烘烤完成後，立即把烤盤取出，把麵包放在放涼架上放涼。

YOMOGI BUNS

/艾草黑糖紅豆麵包/絕配的組合，美食，源於心思

黑糖夾心

黑糖……80克	40%	水……80克 40%
高筋麵粉……20克	10%	粟粉……5克 2.5%
蛋白……28克	14%	牛油……6克 3%

製作黑糖夾心

牛油回復室溫。
黑糖和水放進容器中，浸泡約1.5小時至黑糖完全溶解於水中。
高筋麵粉和粟粉混合，過篩，分2次加入蛋白，以膠刮拌至無粉粒。加入2湯匙黑糖水，攪拌均勻，再分2次加入餘下的黑糖水，拌勻。放進微波爐800w，加熱40秒，取出用膠刮攪拌均勻，再以800w加熱40秒，取出加入牛油拌勻。
用保鮮紙包着再碾成10x10cm片狀，冷藏備用。

麵糰

高筋麵粉......200克　100%
上白糖......20克　10%
鹽......4克　2%
全蛋......50克　25%
蛋黃......20克　10%
原種......60克　30%
牛奶......90克　45%
牛油......20克　10%
艾草......16克　8%

摺入用

無糖紅豆粒......180克　90%

灑面用

黃豆粉......適量
糖粉......相當於黃豆粉的5-15%（因應個人喜好增減）

糖粉過篩後加入黃豆粉拌勻，備用。

製作麵糰

牛油放在室溫至16-18℃。

原種先放在牛奶中浸泡2-3分鐘，以打蛋器攪拌均勻後放進攪拌機之大盆內，放入所有材料（牛油除外），以慢速打至混合，再以高速打至麵糰表面光滑。加入牛油，以中速打至混合及柔滑狀態。

攪拌完成的溫度為23-24℃

用手把麵糰檢查，延展至薄如透明的薄膜即可，把麵糰塑成圓球狀，放入膠盆中。

盆麵糰放在28℃室溫中進行第一次發酵，至膨脹率達1.5倍大，需時約2小時。

乎情況作出調整

麵

麵糰由膠盆倒出，把麵糰四邊往中央摺，再放進膠盆中。

麵可使麵糰交換空氣，增加體積，並使表面更有彈性和緊致。

溫發酵

盆麵糰放進4-6℃雪櫃中，低溫發酵12-24小時至膨脹率達2.5倍。

溫

雪櫃取出整盆麵糰，放在28℃室溫中回溫0.5-1小時。

從雪櫃取時未完成發酵，可放在室溫至膨脹率達2.5倍為止。

造型

用手輕壓麵糰排出空氣，把麵糰拉成四角形。用棍將麵糰碾開成厚度一致的正方形，放入黑糖夾心。將麵糰的四個角摺起，麵糰輕微重疊，並且把接縫處黏合。

用棍將麵糰碾開成約1cm厚的長方形，在2/3的麵糰放上紅豆粒，摺成三摺，用棍輕輕碾過使麵糰貼合。用棍將麵糰碾開成約1cm厚的長方形，切割約2cm濶的長條，共12等分。將麵糰拉長，扭轉數次，隨意打結，排放在烤盤上，噴上水氣。

第二次發酵

麵糰放在28℃的溫度下約1小時，完成最後發酵後，在麵糰表面撒上黃豆粉。

多撒一點黃豆粉覆蓋麵糰表面的紅豆，可防止紅豆烘烤至乾硬。

烘烤

放進已預熱230℃的焗爐，再把焗爐溫度設定為200℃，烘烤12-14分鐘。烘烤完成後，立即把烤盤取出，把麵包放在架上待涼。

設計這包點源於喜歡吃和果子，把草餅變成和果子包點，每一口也嘗到黑糖、艾草、黃豆粉和紅豆的味道。

KOUGLOF

/咕咕洛夫/歐洲的傳統點心，溢出濃濃的酒漬乾果香

模具：直徑13cm　可做6個

酒漬乾燥水果

紅莓乾......30克　20%
糖漬橙皮......30克　20%
無花果乾......30克　20%
藍莓乾......30克　20%
上白糖......30克　20%
金萬利橙酒......90克　60%
水......30克　20%
*可因應個人喜好使用朗姆酒代替金萬利橙酒

自我分解麵糰

法國麵包專用粉......120克　80%
高筋麵粉......30克　20%
上白糖......38克　25%
蛋黃......45克　30%
淡忌廉......30克　20%
牛奶......30克　20%

主麵糰

原種......45克　30%
鹽......3克　2%
牛油......60克　40%

混合用

開心果......45克　30%

模具用

生杏仁......適量

製作酒漬乾燥水果

把所有材料放進鍋中，以中火加熱至沸騰後調至小火煮10分鐘，放進玻璃瓶，待冷卻後放進雪櫃，冷藏浸泡2-5天；使用前放在網篩上1小時，讓酒漬乾燥水果完全瀝乾。

製作自我分解麵糰

把所有材料放入攪拌機之大盆內，攪拌至混合均勻，

把麵糰連同大盆包上保鮮紙，放進4-6℃雪櫃中，冷藏30-60分鐘。

製作主麵糰

牛油回溫至16-18℃。

原種和撕成小塊的自我分解麵糰放入攪拌機之大盆內，加入鹽，以慢速打至混合，再以高速打至麵糰表面光滑，加入一半份量的牛油，以中速攪打至大致混合，再加入餘下的牛油，以中速攪打至混合及柔滑狀態。

加入開心果和已瀝乾的酒漬乾燥水果，以中速攪打至大致混合，用手檢查麵糰，延展至薄如透明的薄膜即可。

攪拌完成的溫度為22-23℃

把麵糰塑成圓球狀，放入膠盆中。

第一次發酵

放在28℃室溫中進行第一次發酵至膨脹率達2倍大，約3小時。

視乎情況作出調整

排氣。冷卻

將麵糰按壓排氣，放在鋪上烘焙紙的盒內，再包上保鮮紙，放進-10℃的冰箱冷藏3小時以上。

3天內使用，使用前放在4-6℃雪櫃中退冰回溫即可。

分割。鬆弛

用刮板把麵糰準確分割成6份，用手輕壓排出麵糰空氣，滾圓，放在已放有烘焙紙的烤盤上，包上保鮮紙，放進4-6℃雪櫃中，冷藏鬆弛20-30分鐘。

造型

將牛油薄薄的掃在模具中，放入生杏仁，用手輕壓麵糰排出空氣，再次滾圓，手指沾一點牛油，在麵糰中央刺穿，把麵糰接口處朝上，貼合地放進模具中。

第二次發酵

放在28℃的室溫約120分鐘。

烘烤

放進已預熱220℃的焗爐，放進烤盤後把焗爐溫度設定為200℃，烘烤25-30分鐘。烘烤完成後，把麵包脫模後立即掃上溫熱糖水，放在架上待涼。

糖水

上白糖......35克
水......30克
金萬利橙酒......15克

在麵糰烘烤時製作

上白糖和水放進鍋中，以中火加熱至沸騰，熄火加入金萬利橙酒。

伴吃檸檬醬

檸檬皮茸......1個份量
檸檬汁......15克
全蛋......20克
上白糖......15克
牛油......30克

上白糖、檸檬汁和檸檬皮茸放在容器中拌勻，加入全蛋，用打蛋器攪拌均勻，放到鍋中隔水加熱，邊用膠刮攪拌至82℃。把已回溫的牛油放在食物處理器中，分3次加入，攪拌至光滑柔潤。放進容器中，冷藏備用。

伴吃時放到擠袋中，注入麵包中央的凹處。

咕咕洛夫(kouglof)是歐洲節慶或祭典時的傳統代表點心，法國、德國、瑞士、匈牙利等等的國家也有自家的傳統做法，有的做成蛋糕，有的做成麵包，不同的地方做法也略有不同。咕咕洛夫可以是蛋糕也可以是麵包，即使做的是麵包，但口感卻像蛋糕，它外表如蛋糕般精緻，麵糰的材料全是製作蛋糕的基本材料，揉入大量酒漬乾果，組織細緻輕盈，濃郁牛油香味中滲透着橙酒的香氣。

咕咕洛夫單吃美味，添加檸檬醬伴吃是另一種滋味，檸檬的香氣跟咕咕洛夫十分合襯，還可飾成小甜點般精緻。

CHOCOLATE KOUGLOF

/朱古力咕咕洛夫/在濃郁的朱古力中嘗到點點焦糖香

模具：直徑13cm，可做4個

自我分解麵糰

法國麵包專用粉......160克　100%
上白糖......32克　20%
蛋黃......48克　30%
淡忌廉......32克　20%
牛奶......56克　35%

主麵糰

原種......48克　30%
可可粉......13克　8%
鹽......3克　2%
牛油......64克　40%

塗料

牛奶焦糖醬......80克
牛奶焦糖醬做法見P.101

伴吃朱古力醬

65%朱古力......115克
淡忌廉......100克
葡萄糖膠......23克
牛油......23克
開心果碎......適量

製作自我分解麵糰

把所有材料放入攪拌機之大盆內，攪拌至混合均勻，

把麵糰連同大盆包上保鮮紙，放進4-6℃雪櫃中，冷藏30-60分鐘。

自我分解法讓麵糰充分水合形成筋度，製作麵糰時可減少攪拌時間，也可避免麵糰溫度過高，麵糰溫度過高會使牛油溶化，以致完成品較乾，失去應有的濕潤口感。

依照P.114「咕咕洛夫」麵糰的製法、第一次發酵、排氣。冷卻

麵糰的製作圖示

分割。鬆弛

用刮板把麵糰準確分割成4份，用手輕壓把麵糰的空氣排出，滾圓，放在已鋪上烘焙紙的烤盤上，包上保鮮紙，放進4-6℃雪櫃中，冷藏鬆弛20-30分鐘。

造型

將牛油薄薄的掃在模具內。

用手輕壓麵糰排出空氣，用棍將麵糰碾開成長方形，用抹刀平均地塗上約20克牛奶焦糖醬，左右和上端分別預留1cm空位，將麵糰從上方往下捲起，並且把接縫處黏和，收口處朝下。將麵糰的前端和末端稍微重疊，按緊，把麵糰收口處朝下，貼合地放進模具中。

第二次發酵

放在28℃的室溫約120分鐘。

烘烤

放進已預熱220℃的焗爐，放進烤盤後把焗爐溫度設定為200℃，烘烤25-30分鐘，烘烤完成後，把麵包脫模放在架上待涼。

伴吃朱古力醬

牛油回復室溫。

淡忌廉和葡萄糖膠放在鍋中，以中火加熱至沸點，倒入朱古力內，靜止2-3分鐘，用打蛋器攪拌均勻。加入牛油拌勻，將朱古力咕咕洛夫的上半部浸在朱古力醬中，再放上開心果碎。

MAPLE CINNAMON APPLE BUNS

/楓糖肉桂蘋果包/每一口也嘗到肉桂的香

模具：直徑13cm，可做4個

肉桂糖

肉桂粉......6克
上白糖......40克

把材料拌勻即可

楓糖蘋果

楓糖漿......18克
牛油......8克
上白糖......15克
紅蘋果......150克
葡萄乾......20克
肉桂粉......1.5克

紅蘋果洗淨切片。楓糖漿、牛油和上白糖放在鍋子裏，以中火加熱至沸騰後調至小火，加入紅蘋果和葡萄乾，煮至紅蘋果軟身熄火，加入肉桂粉拌勻即可，待涼備用。

發酵種

法國麵包專用粉......52克　20%
原種......7克　2.5%
牛奶......40克　15%

製作發酵種

原種放在牛奶中浸泡2-3分鐘，攪拌均勻後加入法國麵包專用粉拌勻。放置在28℃室溫中發酵3小時，再放進4-6℃雪櫃中冷藏發酵18-24小時。

自我分解麵糰

法國麵包專用粉......78克　30%
高筋麵粉......130克　50%
上白糖......26克　10%
蛋黃......78克　30%
淡忌廉......52克　20%
牛油......52克　20%

製作自我分解麵糰

牛油回溫至16-18℃。

把所有材料放入攪拌機之大盆內，攪拌至混合均勻。

攪拌完成的溫度為21-22℃

把麵糰放進已放有烘焙紙的膠盒內，放進4-6℃雪櫃中冷藏18-24小時。

自我分解法讓麵糰充分水合形成筋度，製作麵糰時可減少攪拌時間，也可避免麵糰溫度過高，麵糰溫度過高會使牛油溶化，以致完成品較乾，失去應有的濕潤口感。

主麵糰

原種......72克　27.5%
牛油......104克　40%
鹽......5克　2%
牛奶......26克　10%

製作麵糰

牛油回溫至16-18℃。

把發酵種和自我分解麵糰撕成小塊放入攪拌機之大盆內，加入原種、鹽和牛奶。以慢速打至混合，再以高速打至麵糰表面光滑。加入一半份量的牛油，以中速打至大致混合。再加入餘下的牛油，以中速打至混合及柔滑狀態。用手把麵糰檢查，延展至薄如透明的薄膜即可。

攪拌完成的溫度為22-23℃

把麵糰塑成圓球狀，放入膠盆中。

排氣。冷卻

將麵糰按壓排氣，放在已放有烘焙紙的烤盤上，包上保鮮紙，放進-10℃的冰箱冷藏3小時以上。

3天內使用，使用前放在4-6℃雪櫃中退冰回溫即可。

分割。鬆弛

用刮板把麵糰準確分割成4份，用手輕壓，把麵糰內的空氣排出，滾圓，放在已放有烘焙紙的烤盤上，包上保鮮紙。放進4-6℃雪櫃中，冷藏鬆弛20-30分鐘。

做型

將牛油薄薄的掃在模具中。用手輕壓麵糰排出空氣，用棍將麵糰碾開成長方形，平均地放上約12克肉桂糖，左右和上端分別預留1cm空位。將麵糰從上方往下捲起，並且把接縫處黏和，收口處朝下，切割成10等分。將麵糰、楓糖蘋果連同汁液交錯地放進模具中。

第二次發酵

放在28℃的室溫約120分鐘。

烘烤

放進已預熱220℃的焗爐，放進烤盤後，把焗爐溫度設定為200℃，烘烤25-30分鐘。烘烤完成後，把麵包脫模放在放涼架上放涼。

肉桂的香，楓糖的甜，與蘋果是perfect match，配上質地像蛋糕的布厘歐麵糰，活像小甜點的包點；在懶洋洋的午後，窩在沙發，濃茶、楓糖肉桂蘋果包在旁，一邊看書，一邊享用；我愛冬日長。

Foreword

This book is a labour of love. From conception and writing recipes, to food styling and photography, I did everything. As this is what I do almost every day, I found the process itself very satisfying and enjoyable.

I'm not a professional when it comes to baking or photography. But I kept on studying and refining my skills because of my immense interest in them. I'm kind of a perfectionist too, so that I got quite hung up on little details, be it the taste or the look of my food. I've been always learning from my mistakes.

I wrote this book to introduce the basics of wild yeast starter to readers who find it difficult. I want to share my successful recipes after many painstaking trials with everyone who likes to make their own bread at home. I want to tell everyone that making quality fine baked goods at home isn't quite nearly as hard as you thought.

There are countless wild yeast starter recipes in different countries. I enjoy experimenting with them as they make every day challenging. I'm thankful that I found interest in cooking, baking and photography. Just a loaf of bread is enough for me to feel happy for a whole day. They add colours and fun to my life.

In search of authentic taste of bread

It's not as difficult as it sounds to grow your own wild yeasts from all-natural ingredients. In fact, I find it easier to make and to use than commercial yeasts. The dough undergoes slow fermentation process under low temperature or in frozen state which allows time for the authentic flavours to develop. The dough also has enough time to fully mature bringing out the best in the yeasts. That's how the authentic taste of bread comes through.

Wild yeasts refer to the natural yeasts that you can grow at home. Wild yeasts are present in almost anything. Thus, you can make a wild yeast culture with many different starters.

There are a smorgasbord of recipes on wild yeast starters. Here I'm including those that are simple, with a high success rate, namely raisin and strawberry starters. The raisin starter gives highly replicable and stable yeasts, regardless of seasons. For a richer sweetness and heady aroma, grow wild yeasts from ripe strawberries when they are In season.

Raisin wild yeast starter

Ingredients

150 g raisins *
300 g water
23 g castor sugar
* Make sure you use raisins that are oil-free. Raisins coated in oil don't ferment. Please read the package carefully.

Growing yeasts

Rinse a glass bottle clean. Boil it in water to sterilize. Wait till the bottle is cold, then put raisins, water and sugar into the bottle. Cover the lid tightly. Leave it in a warm and shady spot away from the sun. Shake the bottle twice a day. After each shake, open the lid to release gases from the culture and introduce fresh air.

Fermentation phases

On day 2, the raisins should have picked up water and swell. Some of them may float. On day 3 and 4, you'd see many tiny bubbles in the water and all raisins should float. After shaking the bottle, you'd hear a hissing sound when you open the lid. You'd see even more bubbles rising and it smells vaguely like alcohol.

As the fermentation process approaches the end on day 5 and 6, you'd see tiny bubbles in the liquid even if you don't shake it. You'd hear the hissing sound of gases coming out from time to time. There will be white precipitate of yeasts on the bottom of the bottle. After shaking the bottle, you'd see a huge amount of bubbles rising when you open the lid. You can also smell a strong scent of alcohol.

The raisin wild yeast culture is ready to use.

Strawberry wild yeast starter

Day 1

Day 2

Day 3

Day 3

Day 4

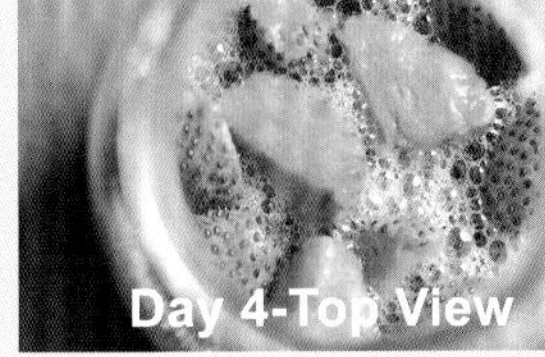
Day 4-Top View

Day 4-White Precipitation

Ingredients

120 g strawberries
300 g water
23 g castor sugar

Growing yeasts

Rinse a glass bottle clean. Boil it in water to sterilize. Wait till the bottle is cold, then put strawberries, water and sugar into the bottle. Cover the lid tightly. Leave it in a warm and shady spot away from the sun. Shake the bottle twice a day. After each shake, open the lid to release gases from the culture and introduce fresh air.

Fermentation phases

On day 2 and 3, the liquid should have turned pink. After shaking the bottle, you'd hear a hissing sound when you open the lid. You'd see bubbles rising and it smells vaguely like strawberries.

As the fermentation process approaches the end on day 4 and 5, you'd see tiny bubbles in the liquid even if you don't shake it. You'd hear the hissing sound of gases coming out from time to time. There will be white precipitate of yeasts on the bottom of the bottle. The strawberries turn pale and float. After shaking the bottle, you'd see a huge amount of bubbles rising when you open the lid. You can also smell a strong scent of strawberries and a light scent of alcohol.

The strawberry wild yeast culture is ready to use.

Storing yeast culture

Pick an airtight glass bottle. Boil it in water to sterilize it. Strain the yeast culture and pour into the bottle.

Cover the lid tightly and store in a fridge.

The yeast culture lasts well for 3 months. Just make sure you add 1 small spoonful of sugar to the culture to keep the yeasts alive.

The yeast culture is then added to flour to make a stable and lively sponge ferment. Your very own wild yeasts are unique and you can feel their liveliness when you handle them.

Yeasts change in different environments, temperature and humidity. There isn't a hard-and-fast formula that guarantees the same result every time. But you may adjust the proportion of water to flour according to the liveliness of the yeast and your personal preferences, for healthy living natural yeast.

Sponge ferment (70% water)

Day 1

In a container, put 50 g wild yeast culture and 50 g wholemeal flour. Mix well. Leave them at room temperature (about 28°C) to ferment for 4 hours until the mixture doubles in size. Keep in the fridge.

Day 2

Take the container out and add 35 g of water and 50 g of wholemeal flour to the mixture. Mix well and leave it to ferment at room temperature until it doubles in size. Keep in the fridge.

Day 3

Take the container out and add 35 g of water and 50 g of wholemeal flour. Mix well. Leave it to ferment at room temperature until it doubles in size. Refrigerate overnight. Use the sponge ferment the next day.

Storing and feeding

You may store the sponge ferment in a fridge infinitely as long as you "feed" it. Whenever you use part of the sponge ferment, you should put in the same weight of water and flour as you the ferment you use. Even if you don't use any, you have to "feed" it once a week to keep the yeasts alive.

For instance, you use 60 g of sponge ferment. You should put back 18 g of water and 42 g of bread flour. Mix well and let the mixture ferment at room temperature until it doubles in size. Then keep in the fridge.

The calculation works like this

60 g X 0.7 (the baker's percentage of water in the dough) = 42 g of bread flour

Then make up the original volume of the sponge ferment with water:

60 g – 42 g = 18 g of water

To "feed" the yeasts, use bread flour and wholemeal flour alternately. If you feed that the yeasts are not very lively, you may use wild yeast culture in place of water.

Baker's percentage

It's a notion to express the ratio of ingredients relative to the weight of flours. The baker's percentage for flours is expressed as 100% by default and other ingredients are expressed as a percentage relative the weight of all flours.

The advantage of this system is that the baker can tell the ratio of the ingredients easily and can foresee the baking results roughly. Once a baker has to change the batch quantity, he can easily follow the percentage for the amount of each ingredient.

Baker's percentage sample calculation

In this example, the original recipe calls for 200 g of flour. A baker has to change the recipe to use 250 g of flour instead.

200 g bread flour, 100%
2 g honey, 1%
6 g castor sugar, 3%
4 g salt, 2%
50 g mashed potato, 25%
70 g water, 35%
10 g olive oil, 5%
40 g sponge ferment, 20%

Changing the recipe to use 250 g of flour

250 g bread flour, 100%	
3 g honey, 1%	< 250 x 1%=2.5 g
8 g castor sugar, 3%	< 250 x 3%=7.5 g
5 g salt, 2%	< 250 x 2%=5 g
63 g mashed potato, 25%	< 250 x 25%=62.5 g
88 g water, 35%	< 250 x 35%=87.5 g
13 g olive oil, 5%	< 250 x 5%=12.5 g
50 g sponge ferment, 20%	< 250 x 20%=50 g

*All weights rounded to the nearest integer

Butter Roll

Rustic and simple, round and cute

Make yourself a sumptuous brunch at home by stuffing the rolls with all kinds of fillings. Slice it in half and sandwich some sliced cucumber and smoked salmon in between. Stuff another with scrambled eggs, grilled mushrooms and smoked ham. Serve them with cream of pumpkin and carrot soup and grilled pumpkin salad. Voila.

(makes 12 rolls)

Ingredients

300 g bread flour, 100%
30 g castor sugar, 10%
4.5 g salt, 1.5%
37.5 g whole eggs, 12.5%
135 g water, 45%
30 g butter, 10%
40 g sponge ferment, 30%

Egg wash

whisked whole eggs

Dough

Warm the butter at room temperature up to 16-18°C.

Soak the sponge ferment in water for 2 to 3 minutes. Beat with whisk until well mixed. Put it into a big mixing bowl of the table-top electric mixer. Add all remaining ingredients (except butter). Knead over low speed with dough hooks until well incorporated. Then knead with high speed until smooth and shiny. Add butter and knead over medium speed until well mixed, soft and smooth.

The dough should be 23 to 24°C after this step.

Check the dough for gluten development by stretching it. If it stretches into a translucent thin membrane, it is good enough. Roll the dough into a sphere. Put it into a plastic tray.

First rise

Leave the dough at room temperature (28°C) for about 2 hours in the First rise. It should expand to 1.5 times of its original size.

Adjust the rising time according to actual conditions

Punch down

Turn the dough out from the plastic tray. Fold the four corners toward the centre. Press it back into the plastic tray.

Punching down the dough helps free up more food for the yeasts and bring in fresh air. It also helps the dough expand further, increasing its elasticity and giving it a finer texture.

Low-temperature rise

Put the tray of dough into a fridge at 4 to 6°C. Leave it to rise at low temperature for 12 to 24 hours. It should expand in size 2.5 times.

Warming up

Take the tray of dough out and leave it at room temperature (28°C) for 30 to 60 minutes. Adjust the time according to actual conditions

If the dough hasn't expanded 2.5 times after staying in the fridge for 24 hours, you may let it complete the fermentation at room temperature until it expands enough.

Dividing / Resting

Cut the dough into 8 equal portions with a dough scraper. Punch down each dough with your hands to press the air out. Round it and cover with damp cloth or a plastic box. Let it rest for 20 to 30 minutes.

Rounding the dough means folding the edge of the dough toward the centre. Put it the seam side down. Then turn the dough with your palms to roll it into a compact sphere.

Shaping

Gently press the dough with a rolling pin to press out the air. Then roll it flat. Fold the top and bottom end toward the centre. Flip the dough with the seam side down. Pinch the seam the secure. Then roll the dough into a cylinder about 12 cm long.

From the wider end, roll the dough out into a rectangle about 22 to 24 cm long, while pulling the dough toward yourself. Roll the dough up from the wider end. Pinch the seam to secure.

Line a baking tray with a baking paper or non-stick silicone mat. (or you may use a non-stick baking tray instead). Arrange the dough pieces on the tray with the seam side down. As the dough will rise and expand further, make sure you leave enough room between them.

Spray water on the dough pieces.

The dough shouldn't be too wet. When you shape the dough, try not to roll too tight either. Otherwise, the rolls won't look chubby and cute as intended. Though the dough doesn't contain much water, the eggs in it help emulsification. As long as the dough rises properly, the rolls will end up fluffy and pillowy after baked.

Second rise (Proofing)

Leave the dough at 28°C for 1 hour for final proofing. It should expand twice as large when final proofing is done. Brush egg wash over the dough pieces.

Egg wash is just whole whisked eggs

Baking

Preheat an oven to 230°C. Put the rolls in the oven and turn to 200°C. Bake for 10 to 15 minutes. Leave the rolls to cool on a wire rack.

Soymilk English Muffin

Light, holey, with a soybean scent

(makes 7 muffins, each 8 cm in diameter, 2.5 cm tall)

Dough

70 g bread flour, 85%
30 g rice flour, 15%
8 g castor sugar, 4%
3 g salt, 1.5%
120 g soymilk, 60%
6 g butter, 3%
60 g sponge ferment, 30%

Garnish

grainy cornmeal

Dough

Warm the butter at room temperature up to 16-18°C.

Soak the sponge ferment in soymilk for 2 to 3 minutes. Beat with whisk until well mixed. Put it into a big mixing bowl of the table-top electric mixer. Add all remaining ingredients (except butter). Knead over low speed with dough hooks until well incorporated. Then knead with high speed until smooth and shiny. Add butter and knead over medium speed until well mixed, soft and smooth.

The dough should be 23 to 24°C after this step.

Check the dough for gluten development by stretching it. If it stretches into a translucent thin membrane, it is good enough. Roll the dough into a sphere. Put it into a plastic tray.

Dividing / Resting

Divide the dough into 8 equal pieces with a dough scraper.

Press each dough with your hand to release the air. Round them. Cover with damp cloth or with a plastic box. Let the dough rest for 20 to 30 minutes.

First rise

Leave the dough at room temperature (28°C) for about 2 hours in the First rise. It should expand to 1.5 times of its original size.

Adjust the rising time according to actual conditions

Punch down

Turn the dough out from the plastic tray. Fold the four corners toward the centre. Press it back into the plastic tray.

Punching down the dough helps free up more food for the yeasts and bring in fresh air. It also helps the dough expand further, increasing its elasticity and giving it a finer texture.

Low-temperature rise

Put the tray of dough into a fridge at 4 to 6°C. Leave it to rise at low temperature for 12 to 24 hours. It should expand in size 2.5 times.

Warming up

Take the tray of dough out and leave it at room temperature (28°C) for 30 to 60 minutes. Adjust the time according to actual conditions

If the dough hasn't expanded 2.5 times after staying in the fridge for 24 hours, you may let it complete the fermentation at room temperature until it expands enough.

Shaping

Press each dough with your hands to drive the air out. Round them again. Spray water on the dough. Dip each in grainy cornmeal. Press it into a round muffin tin that has been greased with butter. Put the filled muffin tins on a baking tray lined with baking paper or non-stick silicone mat.

Spray water on the dough again.

Greasing the muffin tins with melted butter at room temperature makes it easier to turn out the muffins after baked.

Second rise (Proofing)

Leave the whole tray of muffins to proof at room temperature (28°C) for 1 hour until they expand twice in sizes. Cover them with a sheet of baking paper or a non-stick silicone mat. Top it with another baking tray.

Baking

Preheat oven up to 210°C. Put in the tray of muffins and turn to 190°C. Bake for 13 to 18 minutes. Remove from the oven and leave the muffins to cool on a wire rack.

Sandwich Bread

The most basic, rustic baked goods of all

This is a basic recipe with simple steps and ingredients. The loaf is guaranteed to be soft on the inside and crispy on the outside.

You'd hear the loaf giving crackling sound after it is baked. It's because its crust cracks slowly to give it a fluffy crispy texture. Sandwich bread is very versatile. You can make toastie sandwich with it and serve it with spinach, tomato, burdock and toast on the side. It's a quick and sumptuous breakfast.

Dough

280 g bread flour, 100%
182 g milk, 65%
28 g whipping cream, 10%
11 g castor sugar, 4%
4 g salt, 1.5%
11 g butter, 4%
84 g sponge ferment, 30%

Dough

Leave the butter at room temperature till it reaches 16 to 18°C.

Soak the sponge ferment in milk for 2 to 3 minutes. Whisk it well and then transfer into the mixing bowl of a stand mixer with dough hook attached. Put in all ingredients except butter. Beat over low speed to mix well. Then beat over high speed into smooth dough. Add butter and beat over medium speed into soft dough.

The dough should be at 23 to 24°C after mixing.

Stretch the dough to check gluten development. If you can stretch it into a thin translucent film, it is good. Roll the dough round. Put it into a plastic tray.

First rise

Leave the dough at room temperature (28°C) for about 2 hours in the First rise. It should expand to 1.5 times of its original size.

Adjust the rising time according to actual conditions

Punch down

Turn the dough out from the plastic tray. Fold the four corners toward the centre. Press it back into the plastic tray.

Punching down the dough helps free up more food for the yeasts and bring in fresh air. It also helps the dough expand further, increasing its elasticity and giving it a finer texture.

Low-temperature rise

Put the tray of dough into a fridge at 4 to 6°C. Leave it to rise at low temperature for 12 to 24 hours. It should expand in size 2.5 times.

Warming up

Take the tray of dough out and leave it at room temperature (28°C) for 30 to 60 minutes. Adjust the time according to actual conditions

If the dough hasn't expanded 2.5 times after staying in the fridge for 24 hours, you may let it complete the fermentation at room temperature until it expands enough.

Dividing / Resting

Cut the dough into 3 equal portions with a dough scraper. Punch down each dough with your hands to press the air out. Round it and cover with damp cloth or a plastic box. Let it rest for 20 to 30 minutes.

Shaping

Gently press the dough with your hands to press out the air. Then roll it out into a rectangle with a rolling pin. Fold the short ends toward the centre. Pat it with your palms to compact it. Roll it up and pinch the seam to secure. Put all three pieces of dough into the loaf bread tin. Spray water on the dough pieces.

Second rise (Proofing)

Leave the dough in the tin at 28°C for 1 hour for final proofing. It should expand twice as large when final proofing is done. Brush egg wash over the dough pieces.

Baking

Preheat an oven to 230°C. Put the loaf in the oven and turn to 200°C. Bake for 30 to 35 minutes. Take it out immediately when time is up. Turn the loaf out to cool on a wire rack.

Hamburger Bun

A nutritious American lunch with a touch of Japanese flair...

Just slice the bun horizontally and sandwich some lettuce, sliced tomato and home-made wagyu burger steak in between. Drizzle with teriyaki sauce. Serve with deep-fried onion rings and potato wedges on the side. Simply irresistible.

(makes 8 hamburger buns, about 8 cm in diameter and 2.5 cm in height)

Dough

160 g bread flour, 80%
40 g French white flour Type 55, 20%
20 g castor sugar, 10%
3 g salt, 1.5%
20 g whole egg, 10%
128 g water, 64%
16 g butter, 8%
16 g sponge ferment, 30%

Garnish and egg wash

white sesames
whole egg (whisked)

Dough

Let butter warm up at room temperature up to 16 to 18°C.

Soak the sponge ferment in water for 2 to 3 minutes. Whisk until well combined. Then transfer into the mixing bowl of a stand mixer with dough hooks attached. Put in all ingredients except butter. Beat over low speed to mix well. Then beat with high speed into smooth dough. Add butter. Beat over medium speed to mix well into soft dough.

The dough should be at 23 to 24°C after mixing.

Stretch the dough to check gluten development. If you can stretch the dough into a thin translucent film, it is good. Roll the dough into a ball. Put into a plastic tray.

First rise

Put the dough in tray at 28°C for the First rise until it expands to 1.5 times of its original size. It takes about 2 hours.

Adjust the rising time according to the actual situation.

Punch down

Turn the dough out from the plastic tray. Fold the four corners toward the centre. Press it back into the plastic tray.

Punching down the dough helps free up more food for the yeasts and bring in fresh air. It also helps the dough expand further, increasing its elasticity and giving it a finer texture.

Low-temperature rise

Put the tray of dough into a fridge at 4 to 6°C. Leave it to rise at low temperature for 12 to 24 hours. It should expand in size 2.5 times.

Warming up

Take the tray of dough out and leave it at room temperature (28°C) for 30 to 60 minutes. Adjust the time according to actual conditions.

If the dough hasn't expanded 2.5 times after staying in the fridge for 24 hours, you may let it complete the fermentation at room temperature until it expands enough.

Dividing / Resting

Cut the dough into 8 equal portions with a dough scraper. Punch down each dough with your hands to press the air out. Round it and cover with damp cloth or a plastic box. Let it rest for 20 to 30 minutes.

Shaping

Press the dough with your hands to press out the air. Roll it round again. Put it into an 8-cm round tin greased with butter beforehand. Gently press it in and spray water on it. Greasing the tin with butter makes it easier to turn out after baked.

Second rise (Proofing)

Leave the dough at 28°C for 1 hour for final proofing. It should expand twice as large when final proofing is done. Brush egg wash over the dough and sprinkle with white sesames.

Egg wash is just whole whisked eggs

Baking

Preheat an oven to 210°C. Put the buns in the oven and turn to 190°C. Bake for 12 to 15 minutes. Take the buns out immediate when the time is up. Turn them out to cool on a wire rack.

Provençale Fougasse

Rustic flavours from Southern France

Fougasse is a classic bread from Provence, France. It is usually shaped like a leaf, but you may also use your imagination to improvise your signature shapes.

Simply flavoured with herbs, this is a rustic bread with no pretension. Fougasse is crispy in texture and bursts with the sweetness of wholemeal flour. You may pair it with camembert cheese grilled with garlic and rosemary for a simple, but flavourful treat.

Dough

50 g wholemeal flour, 25%
50 g water, 25%
150 g French white flour Type 55, 75%
40 g sponge ferment, 20%
2 g salt, 1%
90 g water, 45%
12 g olive oil, 6%

Topping

dried mixed herbs

Dough

Mix wholemeal flour with 50 g water. Refrigerate for 12 to 24 hours.

Soak sponge ferment in 90 g water for 2 to 3 minutes. Mix until well incorporated. Stir in wholemeal flour dough. Add French white flour Type 55. Mix well with a rubber spatula. Cover with cling film and leave it at room temperature for 20 minutes.

The dough should be at 20 to 22°C after mixing.

Add salt and press the dough with your hands to spread it out evenly. Wet your hands and fold the dough from the edge toward the centre. Cover with cling film and leave it at room temperature for 20 minutes. Add olive oil and fold the edge of the dough toward the centre repeatedly until the oil is incorporated. Cover with cling film. Leave it at room temperature for 30 minutes.

Wet your hands. Fold the edge of the dough toward the centre. Cover with cling film. Put it in the fridge at 4 to 6°C to undergo low-temperature rise for 12 to 24 hours, until it expands by 2.5 times.

Warming up

Take the dough out from the fridge. Leave it at 28°C for 30 minutes to 1 hour. You may adjust the time according to the situation.

If the dough hasn't finished the rise, you can keep rising it at room temperature until it expands in size by 2.5 times.

Shaping

Take the dough out of the plastic box. Pull into a rectangle. Cut the dough into 4 triangles with a dough scraper. Fold the dough in half and insert your finger into the dough. Pull it apart into a triangle about 0.6 cm thick. Put the dough on non-stick silicone mat. Pour in olive oil and make incisions on the surface by cutting at an angle. Gently pull the opening apart and sprinkle with dried herbs.

Second rise (Proofing)

Leave the shaped dough at 28°C for 45 minutes for final proofing.

Baking

Preheat an oven to 250°C. Put the dough in the oven and turn to 220°C. Bake for 8 to 12 minutes. Remove from oven immediately when the time is up. Leave the fougasse to cool on a wire rack.

Pesto Monkey Bread

Herby and tangy, just a bite is enough to wow your taste buds.

Tear yourself a piece of bread and put black tomato, mozzarella, pesto on it. Grate some lemon zest over. Drizzle with lemon juice and olive oil. This is the taste of an herb garden in summer.

Baking pan: 20 cm (L) X 7.5 cm (W) X 6.5 cm (H)

Dough

90 g French white flour Type 55, 50%
90 g bread flour, 50%
3 g salt, 1.5%
117 g water, 65%
36 g sponge ferment, 20%
5 g olive oil, 3%

Pesto

40 g fresh sweet basil leaves
40 g toasted cashews
20 g parmesan cheese
40 g extra-virgin olive oil
1 clove garlic

Pesto

Rinse the basil leaves and wipe them dry. Make sure there's no water on them whatsoever (or the pesto will turn stale easily). Put all ingredients into a blender. Blend until fine. Transfer the pesto into a sterilized glass bottle. Cover the lid and store in the fridge.

You can make a big batch of pesto and save the leftover for later use. It lasts in the fridge for 2 weeks if you pour some olive oil over the pesto to isolate it from the air. It also tends to keep its bright green colour this way. Alternatively, you can freeze it and it lasts in the freezer for months. Pesto tastes great as a pasta sauce and as a topping for bread. You may alter the proportion of the ingredients according to your preference. You may also use pine nuts instead of cashews.

Dough

Mix bread flour with half of the water. Refrigerate for 12 to 24 hours.

Soak sponge ferment in the rest of water for 2 to 3 minutes. Mix until well incorporated. Stir in bread flour dough. Add French white flour Type 55. Mix well with a rubber spatula. Cover with cling film and leave it at room temperature for 20 minutes.

The dough should be at 20 to 22°C after mixing.

Add salt and press the dough with your hands to spread it out evenly. Wet your hands and fold the dough from the edge toward the centre. Cover with cling film and leave it at room temperature for 20 minutes. Add olive oil and fold the edge of the dough toward the centre repeatedly until the oil is incorporated. Cover with cling film. Leave it at room temperature for 30 minutes.

Wet your hands. Fold the edge of the dough toward the centre. Cover with cling film. Put it in the fridge at 4 to 6°C to undergo low-temperature rise for 12 to 24 hours, until it expands by 2.5 times.

Shaping

Take the dough out of the plastic box. Pull into a square. Spread pesto over it evenly. Poke the dough with your fingers until it is about 1 cm thick. Cut the dough into five equal pieces with a dough scraper. Stack the dough on each other. Cut it according to the width of the baking pan. Put the dough into the pan and level the top with your hand.

Second rise (Proofing)

Leave the dough in the pan at 28°C for 30 minutes for final proofing.

Baking

Preheat an oven to 230°C. Put the dough in the oven and turn to 200°C. Bake for 25 to 30 minutes. Remove from oven immediately when the time is up. Leave the bread to cool on a wire rack.

Focaccia

Rosemary, black olives, a taste of the Mediterranean

Focaccia is a classic Italian flat bread. It's light and crispy on the outside and fluffy on the inside with beautiful airy holes in varying sizes. You may put your favourite ingredients on the basic dough, such as vegetables, cheeses, herbs, sun-dried tomatoes in oil, etc. Just let your imagination run free and create your very own focaccia.

I personally want my focaccia with simply black olives and rosemary, to heighten the unique flavours of the dough. This is a classic Mediterranean recipe. Adding mashed potato to the dough helps retain moisture content. Make sure you add mashed potato and olive oil after you knead the dough until smooth because the authentic taste of the olive oil can be retained that way. This is also the crucial factor determining the aroma and taste of the bread.

Dough

200 g bread flour, 100%
2 g honey, 1%
6 g castor sugar, 3 %
4 g salt, 2 %
50 g mashed potato, 25%
70 g water, 35%
10 g olive oil, 5%
40 g sponge ferment, 20%

Topping

shaved parmesan cheese
black olives
olive oil
rosemary

Dough

Rinse the potatoes and steam them with skin on until done. Peel and mash with a fork. Leave it to cool and refrigerate for later use.

Soak the sponge ferment in water for 2 to 3 minutes. Beat with whisk until well mixed. Put it into a big mixing bowl of the table-top electric mixer. Add all remaining ingredients (except olive oil and mashed potato). Knead over low speed with dough hooks until well incorporated. Then knead with high speed until smooth and shiny.

Add olive oil to the mashed potato. Mix well. Add the mixture to the dough. Knead over medium speed until smooth and well mixed. Check the dough for gluten formation with your hands. If the dough can be stretched into a thin translucent membrane, it is good enough. Shape the dough into a sphere. Leave it in a plastic tray.

The dough should be about 23 to 24°C after kneading.

I used an electric mixer to knead the dough in this recipe, but you can also do it with hands. Please refer to p.148 for the recipe of Fougasse.

First rise

Leave the dough at 28°C for about 2 hours in the First rise. It should expand to 1.5 times of its original size.

Adjust the rising time according to actual conditions.

Punch down

Turn the dough out from the plastic tray. Fold the four corners toward the centre. Press it back into the plastic tray.

Punching down the dough helps free up more food for the yeasts and bring in fresh air. It also helps the dough expand further, increasing its elasticity and giving it a finer texture.

Low-temperature rise

Put the tray of dough into a fridge at 4 to 6°C. Leave it to rise at low temperature for 12 to 24 hours. It should expand in size 2.5 times.

Warming up

Take the tray of dough out and leave it at room temperature (28°C) for 30 to 60 minutes.

Adjust the time according to actual conditions.

If the dough hasn't expanded 2.5 times after staying in the fridge for 24 hours, you may let it complete the fermentation at room temperature until it expands enough.

Shaping

Turn the dough out on a counter. Cover it with damp cloth or a plastic box. Let it rest for 20 to 30 minutes. Put the dough on a non-stick mat. Pour olive oil over it. Pat it down with your hands into a rectangle about 1 cm thick.

Second rise (Proofing)

Leave the dough at 28°C for 30 to 40 minutes for final proofing. Brush olive oil on it. Press your fingers into the dough randomly to create indentations. Arrange herbs, black olive and parmesan cheese over it.

Baking

Preheat an oven to 280°C. Transfer the dough onto a baking tray together with the non-stick mat. Put it in the oven and turn to 250°C. Bake for 15 to 20 minutes. Leave the focaccia to cool on a wire rack.

Flour Tortilla

Wash it down with a shot of tequila and feel the unique Latin American vibe.

You can make flour tortillas at home from scratch without breaking a sweat. As I use wholemeal flour in this recipe, the tortillas have a unique rustic smell of wheat to them. They are soft and nice.

You can make tortilla rolls by putting your favourite ingredients on them, such as salsa. Then roll it up for a light refreshing meal. Yummy!

(makes 6 tortillas)

Dough

75 g bread flour, 50%
75 g wholemeal flour, 50%
1.5 g salt, 1%
53 g water, 35%
23 g sponge ferment, 15%

Dough

Mix wholemeal flour with water. Refrigerate for 12 to 24 hours. Transfer into a mixing bowl and add all remaining ingredients. Knead with your hands to mix well. Cover with cling film and leave it at room temperature for 30 minutes. Fold the edge of the dough toward the centre.

Fold it in half. Cover in cling film. Leave it to rest at room temperature for 1 hour.

You may also refrigerate the dough at this point and divide it and shape it the next day.

Dividing / Shaping

Divide the dough into 6 equal pieces with a dough scraper. Round them and leave them rest for 20 to 30 minutes. Roll each dough piece out into a thin disc about 20 cm in diameter.

Frying

Plain Bagel

Dense, chewy and doughy with rich flavours

Bagel has a dense doughy texture and it is chewier than most bread. The long fermentation time brings out the sweetness and rich flavour of the flour.

I especially love its light fluffy skin. They give a crackling sound after baked as its crust cracks and I derive much satisfaction from this sound. Boiling the dough in honey syrup and baking it over high heat for a short period of time help those cracks build up.

To serve, just slice the bagel horizontally in half. Sandwich butter lettuce, egg mayo and smoked salmon in between. This is a great home-made breakfast that you can enjoy every day.

(make 6 plain bagels)

Dough

135 g bread flour, 45%
150 g French white flour Type 55, 50%
15 g wholemeal flour, 5%
15 g castor sugar, 5%
4.5 g salt, 1.5%
135 g water, 45%
75 g sponge ferment, 25%

For boiling the bagels

500 g water
15 g honey

Dough

Soak the sponge ferment in water for 2 to 3 minutes. Whisk to mix well. Transfer into the mixing bowl of a stand mixer with dough hooks attached. Beat over low speed to mix all ingredients. Then turn to medium speed and beat to make the dough shiny and smooth. Check the gluten development by stretching a piece of dough with your hands. If it can stretch into a thin translucent film, it's good. Roll it into a ball and keep in a plastic box. Cover the lid.

The dough should be at 24 to 25°C after kneaded.

First rise

Leave the dough at room temperature (28°C) for about 2 hours in the First rise. It should expand to 1.5 times of its original size.

Adjust the rising time according to actual conditions.

Low-temperature rise

Put the tray of dough into a fridge at 4 to 6°C. Leave it to rise at low temperature for 12 to 24 hours. It should expand in size 2.5 times.

Warming up

Take the tray of dough out and leave it at room temperature (28°C) for 30 to 60 minutes. Adjust the time according to actual conditions.

If the dough hasn't expanded 2.5 times after staying in the fridge for 24 hours, you may let it complete the fermentation at room temperature until it expands enough.

Dividing / Resting

Cut the dough into 6 equal portions with a dough scraper. Punch down each dough with your hands to press the air out. Round it and cover with damp cloth or a plastic box. Let it rest for 20 to 30 minutes.

Shaping

Press each dough with your palms to press out the air. Roll it out into a rectangle with a rolling pin. Roll it from one end to the other. Pinch firmly on the seam to secure. Then roll it into a long cylinder with your palm. Roll the last 1 cm of one end flat with a rolling pin. Twist the dough twice. Wrap the flatten end into the other end. Pinch to seal the seam. Spray water on them. There's no need to rise the second time. You can boil it right away.

Boiling

Put water and honey in a pot. Bring to the boil over high heat. Turn to medium heat and put the dough into the pot. Boil each side of the dough for 20 seconds. Remove from the hot syrup with a strainer ladle. Drain well. Put the dough onto a baking tray with the seam facing down.

The dough will rise and expand in the baking process. Make sure you leave enough space between them.

Baking

Bake the bagels in a preheat oven at 230°C. Turn the temperature down to 200°C. Bake for 15 to 20 minutes.

Chocolate Bagel

Decadent chocolate richness

Making the plain dough first before adding chocolate and cocoa powder gives this chocolate bagel some beautiful marbling and a richer taste.

Dough

150 g bread flour, 50%
150 g French white flour Type 55, 50%
24 g castor sugar, 8%
4.5 g salt, 1.5%
165 g water, 55%
75 g sponge ferment, 25%

Chocolate marbling

45 g 65% cocoa chocolate, 15%
15 g cocoa powder, 5%

For boiling the bagels

500 g water
15 g castor sugar

Dough

Please refer to the method of Plain Bagel on P.158.

Chop up the chocolate. Sprinkle cocoa powder and chocolate over the plain dough. Cut the dough into pieces with a dough scraper. Stack them up and press well. Repeat cutting and stacking until the chocolate and cocoa are well distributed within the dough.

Bagel dough is dryer than regular bread dough. In case the dough gets too dry, you may spray water on it before kneading. Of course, you may also put the chopped chocolate and cocoa powder into a stand mixer together with other dry ingredients.

Feel free to do it your own way.

Please refer to P.158 for First rise, low temperature rise, warming up, dividing/resting, shaping, boiling and baking steps in the Plain Bagel recipe.

French Toast Bagel

A sumptuous variation of the chocolate bagel

French toast doesn't have to be made with sandwich bread. You can actually use any bread you like.

I always use stale bread that has been lying around for days for French toast. Fresh bread tends to pick up too much egg and the French toast ends up too moist and soggy. If you use fresh bread, just dip it for a short while in the egg mixture.

I used chocolate bagel for this recipe. I store it in a plastic box and refrigerate it after dunking it in egg mixture. I found this step ensures the French toast won't be too soggy after baked. You may also experiment according to your own taste. Refrigerating the bagel with the egg mixture in the fridge tends to give you a moist French toast.

In this recipe, I baked the French toast so that it would be nicely browned and crisped up. You may also fry it in a pan the traditional way until both sides golden. I serve it with creamy cheese, chocolate sauce and berry sauce as a dessert course.

Ingredients

3 chocolate bagels

Egg coating

50 g eggs
25 g icing sugar
12 g honey
50 ml milk
50 g whipping cream

Berry sauce

50 g strawberries
50 g raspberries
50 g cherries
50 g blueberries
15 g Sanontou (Japanese fine brown sugar)
15 g sugar
10 g lemon juice

Chocolate sauce

20 g 65% cocoa chocolate
8 g milk

Cloth-wrapped cheese

50 g cream cheese
17 g sugar
3 g lemon juice
50 g plain yoghurt
50 g whipping cream

Egg coating

Sieve the icing sugar into a mixing bowl. Add eggs and honey. Beat with a whisk. Slowly add milk and cream. Then whisk until well mixed.

French toast bagel

Cut each chocolate bagels into six pieces. Dunk them into the egg mixture. Make sure they suck in the egg mixture on both sides. Put them in an airtight storage box and refrigerate for 12 to 24 hours. Arrange the bagel pieces on a baking tray lined with baking paper or non-stick silicone mat. Bake in a preheated oven at 230°C. Then turn the temperature down to 200°C immediately. Bake for 10 to 15 minutes. Remove from the oven immediately when the time is up. Leave the bagels to cool on a wire rack.

Berry sauce

Dice strawberries. Put berries and cherries into a pot. Add Sanontou, sugar and lemon juice. Cook over medium heat until berries and cherries are soft. Turn off the heat and set aside.

Chocolate sauce

Melt the chocolate over a pot of simmering water. Heat up the milk over a pot of simmering water. Pour milk into the chocolate and stir to mix well.

Cloth-wrapped cheese

Beat the whipping cream with an electric mixer until soft peaks form. Warm the cream cheese up to room temperature. Put cream cheese into a mixing bowl. Add sugar and beat with an electric mixer until well combined. Put in 1/3 of the yoghurt at a time and beat well after each addition. Add lemon juice and mix well with a rubber spatula. Add cream at last and fold well.

Line a round ramekin with paper towel (folded into a square). Then line it with cheese cloth. Pour in the cheese mixture. Fold the cheese cloth up and secure with a rubber band or twist tie. Refrigerate for at least 12 hours. Remove the cheese cloth before serving.

Croissant

Make sure you wipe the crumbs off your face if you don't want to share

The croissants should show clear layers of dough and fluffy honeycomb structure when sliced. They should be crispy and flaky on the outside, but moist and chewy on the inside. All in all, each bite oozes the buttery richness which is the highlight of each croissant.

When you bake the croissants, make sure they are baked through. They can be a bit darker on the outsides, as long as they aren't burnt. Those under-baked croissants are too light in colour and they are too soggy without the crispy texture and buttery aroma.

(makes 13 croissants)

Dough

200 g French white flour Type 55, 100%
10 g castor sugar, 5%
4 g salt, 2%
60 g water, 20%
50 g milk, 25%
60 g sponge ferment, 30%
10 g butter, 5%

Other ingredients

130 g cold butter for laminating, 65%
1 beaten egg (as egg wash)

Dough

Put sponge ferment, milk and water into the stand mixer. Beat well. Put in the remaining ingredients and beat on low speed with dough hooks attached. Then turn to medium speed and beat until dough is smooth.

The dough should be at 20 to 21°C after mixed.

Wrap the dough in cling film. Roll it out into a square about 1 cm thick with a rolling pin. Put it in a freezer at -10°C for 3 to 4 hours. Then transfer to refrigerator at 5°C and leave it there for 8 hours.

Cold butter for laminating

Let butter warm up until pliable, but not too soft.

Wrap it in cling film. Roll it into a square about 0.5 cm thick. Refrigerate for later use.

Laminating the dough / Resting

Remove the cold butter from the fridge. Let it warm up at room temperature until just pliable.

Roll the dough out into a square of equal thickness. Put the slab of butter over it with the corners pointing at the mid-point of each side of the dough. Fold the 4 corners of the dough up to cover the butter like an envelope. Make sure the seams overlap. Pinch the seam to secure well.

Make sure you don't seal in any air when you wrap the butter with dough.

Roll the stuffed dough out until 0.3 to 0.5 cm thick. Trim off the edges (about 0.5 cm wide) with a wheel cutter. Fold in thirds. Roll with a rolling pin to make the layers adhere to each other well. Wrap the dough in cling film. Refrigerate for 30 minutes to 1 hour to rest the dough. Repeat this step twice, meaning the dough is folded in thirds for three times at last.

Shaping

Take the dough out of the fridge. Roll it out into a long rectangle about 2.5 mm thick. Trim off the edges with a wheel cutter (about 0.5 cm wide). With the wheel cutter, cut the dough into isosceles triangles, with the base measuring 9 cm and the height measuring 22 cm. Wet the pointy tip slightly with water. Roll the base up into croissants.

Wetting the tip of the dough helps secure the seam so that the croissants won't come loose in the baking process.

Put the croissants with the seam facing down onto a baking tray lined with baking paper. Spray water over the croissant.

The croissants will expand when heated. Make sure you leave enough space between them.

You may line the baking tray with baking paper, silicone baking mat, or use a non-stick cookie sheet.

Second rise

Leave the croissants at 28°C for 120 minutes until they expand to twice their original size. Brush egg wash over them.

Baking

Preheat an oven to 250°C. Put in the tray of croissants and immediately turn down the oven to 220°C. Bake for 12 to 13 minutes. Remove the croissants from the oven immediately when the time is up. Transfer the croissants onto a wire rack and let cool.

Black Sesame Croissant

Nutty sesames meet buttery richness. How can you resist?

(makes 13 croissants)

Dough

200 g French white flour Type 55, 100%
32 g black sesames, 16%
10 g castor sugar, 5%
4 g salt, 2%
100 g milk, 50%
10 g butter, 5%
60 g sponge ferment, 30%

Other ingredients

130 g cold butter for laminating, 65%
1 beaten egg (as egg wash)

Dough

Please refer to the recipe of plain croissant on p.164 for method.

Cold butter for laminating: please refer to the recipe of Plain croissant on p.164 for method.

Laminating the dough / Resting

Remove the cold butter from the fridge. Let it warm up at room temperature until just pliable.

Roll the dough out into a square of equal thickness. Put the slab of butter over it with the corners pointing at the mid-point of each side of the dough. Fold the 4 corners of the dough up to cover the butter like an envelope. Make sure the seams overlap. Pinch the seam to secure well.

Make sure you don't seal in any air when you wrap the butter with dough.

Roll the stuffed dough out until 0.3 to 0.5 cm thick. Trim off the edges (about 0.5 cm wide) with a wheel cutter. Fold in quarters towards the centre. Roll with a rolling pin to make the layers adhere to each other well. Wrap the dough in cling film. Refrigerate for 30 minutes to 1 hour to rest the dough. Repeat this step once, meaning the dough is folded in quarters for twice at last.

Please refer to the recipe of plain croissant on p.164 for shaping, Second rise and baking.

Almond Croissant

Turn a croissant into a sweet treat

Crispy and buttery croissants covered in almond crumble with a generous sprinkle of flaked almonds are the perfect match with tiramisu.

Ingredients

4 plain croissants

Almond crumble

25 g butter
16 g icing sugar
60 g ground almond
25 g egg
5 g cake flour

Garnish

flaked almonds

Almond crumble

Let the eggs and butter warm up to room temperature. In a stainless steel bowl, put in butter and icing sugar. Mix well with a rubber spatula. Add dry ingredients and mix again. Put in the whole eggs a little at a time. Whisk well. Refrigerate for later use.

Shaping

Arrange the croissants on a cookie sheet. Pipe the almond crumble mixture evenly on them. Sprinkle with flaked almonds on top.

Baking

Preheat an oven to 230°C. Put in the croissants and turn the oven down to 200°C immediately. Bake for 15 to 20 minutes. Remove the croissants from the oven when the time is up. Put the croissants on a wire rack to let cool.

Strawberry Danish

It might not sound spectacular, but you'd never forget it once you've tried it.

Dough

Please refer to the recipe of Croissant on p.163-164 for ingredients and method.

Custard

50 g milk
10 g egg yolk
10 g sugar
4 g cake flour
1/6 vanilla pod
4 g butter

Custard

Let the butter warm up to room temperature.

Slit open the vanilla pod and scrape out the seeds. In a large mixing bowl, mix egg yolk, sugar and flour with a rubber spatula.

In a pot, put in milk, vanilla seeds and vanilla pod. Cook over medium heat until steam comes out (without boiling). Stir in the egg yolk mixture little by little. Mix well with a rubber spatula after each addition. Pass the mixture through mesh strainer. Put it back into the pot and cook over medium heat until thick.

Make sure you keep stirring the custard with a rubber spatula to avoid burning.

Pass the custard through mesh strainer. Add butter and stir well. Cover the custard with cling film and let the cling film touches the custard (to prevent a crust from forming). Let cool. Refrigerate for 1 to 2 hours.

Shaping

Take the dough out of the fridge. Roll it out until 3 mm thick with a rolling pin. Trim off the edges with a wheel cutting (about 0.5 mm wide). Cut dough into 10 cm x 10 cm squares. Wet all corners with some water. Fold the corners towards to centre. Make sure all corners overlap. Press with your finger to secure.

Wetting the corner helps the dough to adhere well.

Put the Danishes into a 8-cm round tart tin. Spray water over it.

Second rise

Let the Danishes to rise at 28°C for 120 minutes until they double in sizes.

Brush egg wash on the Danishes.

Egg wash is just whole whisked eggs.

Pipe some custard at the centre.

Baking

Preheat an oven to 250°C. Put in the Danishes and turn the oven down to 220°C immediately. Bake for 12 to 13 minutes. Take the Danishes out when time is up. Transfer them onto a wire rack to let cool.

Sprinkle with icing sugar. Top with strawberries.

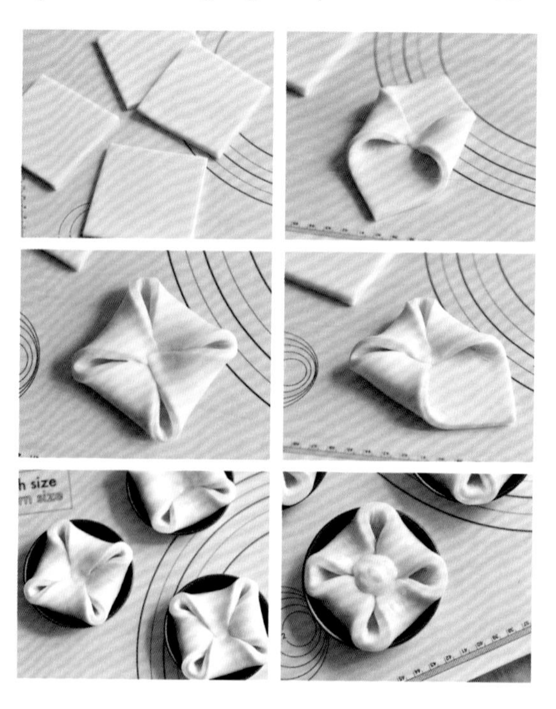

Cinnamon Walnut Buns

Nutty walnut paste with a hint of cinnamon,
alongside coffee icing and caramel nuts. What a treat!

(makes 8 buns)

Dough

180 g bread flour, 100%
18 g castor sugar, 10%
2 g salt, 1%
90 g whole eggs, 50%
54 g butter, 30%
54 g sponge ferment, 30%

Garnish

flaked almonds

Coffee icing

50 g icing sugar
6 g instant coffee
hot water

Almond crumble

35 g butter, 19.4%
35 g whole eggs, 19.4%
35 g icing sugar, 19.4%
15 g cake flour, 8.3%
35 g ground almond, 19.4%

Cinnamon walnut paste

120 g walnuts, 66.6%
20 g butter, 11%
6 g ground cinnamon, 3%
35 g castor sugar, 19.4%

Other ingredients

1 beaten egg (as egg wash)
caramel hazelnuts
caramel almonds

Almond crumble

Let the eggs and butter warm up to room temperature. In a stainless steel bowl, put in butter and icing sugar. Mix well with a rubber spatula. Add dry ingredients and mix again. Put in the whole eggs a little at a time. Whisk well. Refrigerate for later use.

Cinnamon walnut paste

Bake the walnuts in a preheated oven at 170°C for 10 to 15 minutes. Let cool. Let the butter warm up to room temperature. Put all ingredients into a food processor and blend until fine. Wrap the paste in cling film. Roll it out into a thin layer about 0.5 cm thick. Refrigerate for later use.

Dough

Warm the butter at room temperature up to 16-18°C.

Add all ingredients (except butter) in the electric mixer. Knead over low speed with dough hooks until well incorporated. Then knead with high speed until smooth and shiny. Add butter and knead over medium speed until well mixed, soft and smooth.

The dough should be 23 to 24°C after this step.

Check the dough for gluten development by stretching it. If it stretches into a translucent thin membrane, it is good enough. Roll the dough into a sphere. Put it into a plastic tray.

First rise

Leave the dough at room temperature (28°C) for about 2 hours in the First rise. It should expand to 1.5 times of its original size.

Adjust the rising time according to actual conditions

Punch down

Turn the dough out from the plastic tray. Fold the four corners toward the centre. Press it back into the plastic tray.

Punching down the dough helps free up more food for the yeasts and bring in fresh air. It also helps the dough expand further, increasing its elasticity and giving it a finer texture.

Low-temperature rise

Put the tray of dough into a fridge at 4 to 6°C. Leave it to rise at low temperature for 12 to 24 hours. It should expand in size 2.5 times.

Warming up

Take the tray of dough out and leave it at room temperature (28°C) for 30 to 60 minutes. Adjust the time according to actual conditions

If the dough hasn't expanded 2.5 times after staying in the fridge for 24 hours, you may let it complete the fermentation at room temperature until it expands enough.

paste filling on top. Roll the dough from the top down. Pinch the seam to secure. Put it on counter with the seam down. Cut into 8 equal pieces.

Put the marbled dough pieces on a counter with the cut side up. Take the 8 small pieces of plain dough out of the fridge. Press them flat so that they are big enough to cover the marbled dough. Put it on the bottom of each piece of marbled dough. Arrange on a baking tray. Spray water on them. Press each piece flat with your palms.

Add a little hot water to the instant coffee. Stir until it dissolves. Sieve the icing sugar. Add instant coffee mixture to the icing sugar. Mix well. Stir in water little by little with a rubber spatula into a thick paste. Cover the container with damp cloth. Set aside.

Brioche

The French classic that's intensely buttery

Brioche is a sweet pastry with high egg and butter content. It's fluffy and moist with an intense aroma of butter. Serve it with blueberry jam to heighten its flavour.

Preferment

52 g French white flour Type 55, 20%
7 g sponge ferment, 2.5%
40 g milk, 15%

Soak the sponge ferment in milk for 2 to 3 minutes. Stir well. Add flour and mix again. Keep it at 28°C for 3 hours. Then put in a fridge and let it ferment for 18 to 24 hours at 4 to 6°C.

Autolyse dough

78 g French white flour Type 55, 30%
130 g bread flour, 50%
26 g castor sugar, 10%
78 g egg yolk, 30%
52 g whipping cream, 20%
52 g butter, 20%

Let butter warm up to room temperature.

Put all ingredients into a stand mixer. Mix until well combined.

The dough should be at 21 to 22°C when mixing is done.

Put the dough into a plastic box lined with baking paper. Refrigerate at 4 to 6°C for 18 to 24 hours.

The autolyse process lets the flour to combine with water fully for better links between gluten and starches. It shortens the time required to mix the dough while preventing the dough to heat up too much. If the dough is too hot, the butter will melt and be absorbed by the flour. The brioche will turn out dryer in texture, instead of moist and fluffy.

Main dough

72 g sponge ferment, 27.5%
5 g salt, 2%
104 g butter, 40%
26 g milk, 10%

Let the butter warm up to room temperature (about 16 to 18°C).

Put preferment into the mixing bowl of the stand mixer. Tear the autolyse dough into pieces and put them into the same bowl. Add sponge ferment, salt and milk. Beat with dough hooks attached over low speed. Then turn to high speed and beat until smooth. Add half of the butter. Beat over medium speed to mix well. Put in the rest of the butter and beat with medium speed until soft and well combined. Check the dough with windowpane test. It's done if you can stretch the dough into a thin translucent membrane.

The dough should be at 22 to 23°C when the mixing is done.

Roll the dough into a ball. Put into a plastic tray.

First rise

Let the dough sit at 28°C for the First rise until it doubles in size. It takes about 3 to 4 hours.

Adjust the time if needed.

Punch down / Cooling

Press the dough flat to drive the air out. Put it into a baking tray lined with baking paper. Wrap it in cling film and keep it in a freezer at -10°C for at least 3 hours (up to 3 days).

The dough will be frozen if kept in the freezer for too long. In that case, let it sit in the fridge at 4 to 6°C for at least 8 hours before dividing.

Dividing / Resting

Cut the dough into 12 equal parts with a scraper. Gently press each piece with your hands to drive the air out. Roll it round. Put into a baking tray lined with baking paper. Cover with cling film and refrigerate at 4 to 6°C for 20 to 30 minutes to rest them.

Shaping

Press the air out with your hands. Roll it round again. Put 6 pieces of dough into the same bread tin.

Second rise

Leave the brioche at 28°C for 120 minutes until it doubles in size.

Brush egg wash on the brioche.

Egg wash is just whole whisked eggs.

Baking

Preheat an oven to 220°C. Put in the brioche and turn the oven down to 200°C. Bake for 30 to 35 minutes. Remove the brioche from the oven when time is up. Turn the brioche out of the tin and leave it on a wire rack to let cool.

Coffee Monkey Bread

Start your day the perfect way with something sweet and rustic

The rustic buns look cute and unpretentious, exuding a rich coffee aroma. It tastes even better with a dribble of Dulce de leche.

Dough

200 g bread flour, 100%
20 g Sanontou (Japanese fine brown sugar), 10%
2 g salt, 1%
30 g whole eggs, 15%
60 g sponge ferment, 30%
90 g milk, 45%
30 g butter, 15%
40 g dried cranberries, 20%

Coffee syrup

10 g instant coffee, 5%
15 g hot water, 7.5%
25 g butter, 12.5%
25 g Sanontou (Japanese fine brown sugar), 12.5%
30 g almonds, 15%

Other utensils and ingredients:

3 square cake tins (6.5 cm)
3 Pandoro tins (8.5 cm diameter, 6.5 cm deep)
coarsely ground cornmeal

Coffee syrup

Bake the almonds in a preheated oven at 170°C for 8 to 10 minutes. Let cool and finely chop them.

Add instant coffee to the hot water. Stir until it dissolves.

Melt butter over a pot of simmering water or double boiler. Add Sanontou and mix well. Add coffee and mix again. Put in the chopped almonds right before using.

If you make the syrup before shaping the dough, the butter may curd at room temperature. Just heat it up again over a double boiler until it melts before using.

Dough

Warm the butter at room temperature up to 16-18°C.

Soak the sponge ferment in milk for 2 to 3 minutes. Beat with whisk until well mixed. Put it into a big mixing bowl of the table-top electric mixer. Add all remaining ingredients (except butter). Knead over low speed with dough hooks until well incorporated. Then knead with high speed until smooth and shiny. Add butter and knead over medium speed until well mixed, soft and smooth.

The dough should be 23 to 24°C after this step.

Check the dough for gluten development by stretching it. If it stretches into a translucent thin membrane, it is good enough. Roll the dough into a sphere. Put it into a plastic tray.

Soak the cranberries in water for 30 minutes. Drain through a wire strainer and let it sit in the strainer for 10 minutes.

Pull the dough into a square and put it on the counter. Put on the cranberries and roll it up. Punch flat with your palms. Roll it up again. Shape the dough into a ball and transfer into a plastic tray.

First rise

Leave the dough at room temperature (28°C) for about 2 hours in the First rise. It should expand to 1.5 times of its original size.

Adjust the rising time according to actual conditions.

Punch down

Turn the dough out from the plastic tray. Fold the four corners toward the centre. Press it back into the plastic tray.

Punching down the dough helps free up more food for the yeasts and bring in fresh air. It also helps the dough expand further, increasing its elasticity and giving it a finer texture.

Low-temperature rise

Put the tray of dough into a fridge at 4 to 6°C. Leave it to rise at low temperature for 12 to 24 hours. It should expand in size 2.5 times.

Warming up

Take the tray of dough out and leave it at room temperature (28°C) for 30 to 60 minutes. Adjust the time according to actual conditions

If the dough hasn't expanded 2.5 times after staying in the fridge for 24 hours, you may let it complete the fermentation at room temperature until it expands enough.

Dividing

Cut the dough into random pieces with a scraper. Gently press each piece with your hands to drive the air out. Roll them round.

Shaping

Dunk the dough into the coffee syrup. Spray water on it. Roll th e dough in coarsely ground cornmeal. Press it into a cake tin greased with butter. Spray water again.

Greasing the tin with butter helps prevent the buns from sticking to the tin after baked.

Second rise

Let the bun in the tins to rise at 28°C for 60 minutes until they double in sizes. Cover the tin with baking paper or non-stick silicone mat. Top with a baking tray.

Baking

Preheat an oven to 210°C

Put in the buns and turn the oven down to 190°C

Bake for 13 to 18 minutes. Remove the buns from the oven when time is up. Transfer the buns onto a wire rack to let cool.

Dulce de leche is a traditional Spanish confection. "Dulce" means candy and "leche" is milk. Dulce de leche is widely enjoyed in various ways, such as spreading it on toast and cookies, used in desserts, or even stirred in drinks like hot chocolate or coffee. There are also many variations of dulce de leche, by incorporating other ingredients, such as cinnamon, coffee, rum and chocolate.

Dulce de leche is made from condensed milk or milk and sugar. It's easy to make and it's thick and creamy, with golden caramel colour and rich aroma.

Method

Using a steamer

Put a can of condensed milk into a steamer and steam at 100°C for 3 hours. Let it cool down completely in the steamer. Without opening the can, it lasts for 1 month in the fridge.

After opening the can, you can keep the dulce de leche in a glass bottle. It lasts in the fridge for 1 week.

Using a saucepan

Wash glass bottle. Boiling glass bottle for a few minutes, drain and cool well before use. Pour the condensed milk into glass bottle and seal with the cap. Put the glass bottle into a saucepan. Fill the saucepan to one-third with water. Cover the lid. Cook over low heat for 3 to 4 hours. Let it cool down completely in the saucepan.

Without opening the glass bottle, it lasts for 1 month in the fridge. After opening the glass bottle, it lasts in the fridge for 1 week.

Cooking time depends on your stove and the container you use. Please adjust the cooking time according to the texture and the colour of the dulce de leche.

Using an oven

Preheat an oven to 200°C.

Pour the condensed milk into an oven-safe container (preferably one with wide bottom which shortens the baking time). Cover the top with aluminium foil. Put the container into a deep baking tray. Fill the baking tray to half full with boiling hot water. Put in the oven and turn the oven down to 180°C. Bake for at least 90 minutes. Take the condensed milk out every 30 minutes and stir thoroughly with a rubber spatula. Pour the dulce de leche into a glass bottle while still hot. Let cool and refrigerate. It lasts in the fridge for 1 week.

Baking time depends on your oven and the container you use. Please adjust the baking time according to the texture and the colour of the dulce de leche.

Black Sesame Loaf

One bite is never enough

The braiding gives this loaf a lovely marbling pattern with the nutty sesame filling. You may also serve it with some sweetened red bean paste for an extra dimension of flavours.

Dough

140 g bread flour, 100%
17 g Sanontou (Japanese fine brown sugar), 12%
2 g salt, 1.5%
14 g whole eggs, 10%
73 g water, 52%
18 g butter, 13%
42 g sponge ferment, 30%

Black sesame filling

70 g unsweetened black sesame paste, 50%
49 g milk, 35%
30 g black sesames, 22%
10 g sugar, 7%
20 g cake flour, 14%
5 g butter, 3.5%

Leave the butter at room temperature until soft. Sieve the cake flour into a bowl. Add black sesame paste and sugar. Mix well. Add a little milk at a time and mix well with a rubber spatula after each addition.

Heat the dough in a microwave oven at 500W for 40 seconds. Stir with a rubber spatula into a thick paste. Heat it again in a microwave oven at 500W for 40 more seconds. Stir in butter. Wrap in cling film and roll it out into a thin square about 15 cm x 15 cm. Refrigerate for later use.

Dough

Warm the butter at room temperature up to 16 to 18°C.

Soak the sponge ferment in water for 2 to 3 minutes. Beat with whisk until well mixed. Put it into a big mixing bowl of the table-top electric mixer. Add all remaining ingredients (except butter). Knead over low speed with dough hooks until well incorporated. Then knead with high speed until smooth and shiny. Add butter and knead over medium speed until well mixed, soft and smooth.

The dough should be 23 to 24°C after this step.

Check the dough for gluten development by stretching it. If it stretches into a translucent thin membrane, it is good enough. Roll the dough into a sphere. Put it into a plastic tray.

First rise

Leave the dough at room temperature (28°C) for about 2 hours in the First rise. It should expand to 1.5 times of its original size.

Adjust the rising time according to actual conditions

Punch down

Turn the dough out from the plastic tray. Fold the four corners toward the centre. Press it back into the plastic tray.

Punching down the dough helps free up more food for the yeasts and bring in fresh air. It also helps the dough expand further, increasing its elasticity and giving it a finer texture.

Low-temperature rise

Put the tray of dough into a fridge at 4 to 6°C. Leave it to rise at low temperature for 12 to 24 hours. It should expand in size 2.5 times.

Warming up

Take the tray of dough out and leave it at room temperature (28°C) for 30 to 60 minutes. Adjust the time according to actual conditions

If the dough hasn't expanded 2.5 times after staying in the fridge for 24 hours, you may let it complete the fermentation at room temperature until it expands enough.

Laminating the dough / Resting

Press the dough with your palms gently to drive out the air. Pull the dough into a rectangle. Then roll the dough out into a square of equal thickness. Put the black sesame filling on the dough. Fold the four corners of the dough up to cover the filling. Make sure the seams overlap and pinch the seams to secure well.

Roll the dough out until 1 cm thick. Fold toward the centre in quarters. Gently roll the stacked dough to make sure the layers adhere well.

Repeat this step once more.

Refrigerate the dough for 20 to 30 minutes to rest it.

Shaping

Take the dough out of the fridge and roll it out into equal thickness. Cut into 6 equal strips. Braid three of the dough strips together. Roll them up and press into a loaf tin. Spray with water. Repeat with the rest of the dough strips.

Second rise

Leave the loaves at 28°C for 60 minutes until they double in sizes. Cover the lids of the loaf tins.

Baking

Preheat an oven to 230°C. Put in the loaves and turn the oven down to 200°C immediately. Bake for 25 to 30 minutes. Remove the loaves from the oven once the time is up. Turn them out of the loaf tins and let cool on a wire rack.

Yomogi Buns

The perfect union of flavours and texture; good food stems from mindful thoughts

I love Japanese sweets and this recipe is a variation of the traditional Kusa Mochi. You'd get to taste the black sugar, yomogi, kinako and red beans in every bite.

Black sugar filling

80 g black sugar, 40%
80 g water, 40%
20 g bread flour, 10%
5 g cornstarch, 2.5%
28 g egg white, 14%
6 g butter, 3%

Warm the butter up to room temperature.

In a bowl, put in water and black sugar. Leave it for 1.5 hours until the sugar dissolves completely.

Sieve the bread flour and cornstarch together. Add half of the egg white at a time. Stir with a rubber spatula until lump free after each addition. Add 2 tbsp of the sugar syrup. Mix well. Then pour in half of the remaining syrup. Stir well. Pour in the rest of the syrup and stir again. Heat in a microwave at 800W for 40 seconds. Stir with a rubber spatula. Heat the mixture again for 40 seconds at 800W. Add butter and stir well.

Wrap the resulting sugar mixture in cling film. Roll into a square about 10 cm x 10 cm. Refrigerate until firm.

Dough

200 g bread flour, 100%
20 g castor sugar, 10%
4 g salt, 2%
50 g whole eggs, 25%
20 g egg yolk, 10%
60 g sponge ferment, 30%
90 g milk, 45%
20 g butter, 10%
16 g yomogi powder (Japanese mugwort) 8%

Red bean filling

180 g unsweetened cooked red beans, 90%

Garnish

kinako (roasted soybean powder)

icing sugar (about 5 to 15% of kinako by weight, you may adjust the amount to taste)

Sieve the icing sugar and mix well with kinako. Set aside.

Dough

Warm the butter at room temperature up to 16-18°C.

Soak the sponge ferment in milk for 2 to 3 minutes. Beat with whisk until well mixed. Put it into a big mixing bowl of the table-top electric mixer. Add all remaining ingredients (except butter). Knead over low speed with dough hooks until well incorporated. Then knead with high speed until smooth and shiny. Add butter and knead over medium speed until well mixed, soft and smooth.

The dough should be 23 to 24°C after this step.

Check the dough for gluten development by stretching it. If it stretches into a translucent thin membrane, it is good enough. Roll the dough into a sphere. Put it into a plastic tray.

First rise

Leave the dough at room temperature (28°C) for about 2 hours in the First rise. It should expand to 1.5 times of its original size.

Adjust the rising time according to actual conditions

Punch down

Turn the dough out from the plastic tray. Fold the four corners toward the centre. Press it back into the plastic tray.

Punching down the dough helps free up more food for the yeasts and bring in fresh air. It also helps the dough expand further, increasing its elasticity and giving it a finer texture.

Low-temperature rise

Put the tray of dough into a fridge at 4 to 6°C. Leave it to rise at low temperature for 12 to 24 hours. It should expand in size 2.5 times.

Warming up

Take the tray of dough out and leave it at room temperature (28°C) for 30 to 60 minutes. Adjust the time according to actual conditions.

If the dough hasn't expanded 2.5 times after staying in the fridge for 24 hours, you may let it complete the fermentation at room temperature until it expands enough.

Shaping

Press the dough to release the air. Pull the dough into a rectangle. Roll it into a square of equal thickness with a rolling pin. Put the square of black sugar filling at the centre, making a 90 degree angle with the dough. Fold the four corners of the dough toward the centre to wrap the filling. Make sure the dough overlaps a little on every seam. Pinch the dough to seal the seams well.

Roll the stuffed dough out with a rolling pin into a rectangle about 1 cm thick. Spread the red beans on two-thirds of the dough. Fold the dough in thirds by folding the bare third inward twice. Roll with a rolling pin gently to adhere the layers. Then use the rolling pin to roll it out into a rectangle about 1 cm thick. Cut into 12 equal strips, each about 2 cm wide. Pull each string of dough, twist a few times randomly and tie a knot into a bun. Arrange on a baking tray. Spray water on them.

Second rise (Proofing)

Leave the buns at room temperature (28°C) for 1 hour. Sprinkle with sweetened kinako over the dough. Sprinkle kinako generously to cover the red beans. This helps prevent them from drying out after being baked.

Baking

Preheat an oven up to 230°C. Put in the buns and turn to 200°C. Bake for 12 to 14 minutes. Remove from oven and leave the buns to cool on a wire rack.

Kouglof

A traditional European treat with the richness of wine-marinated dried fruit.

Kouglof is a representative festive snack in Europe. Almost all European countries, including France, Germany, Switzerland and Hungary, have their own traditional version of Kouglof. Some of them are more like cake. Some of them are more like bread. The recipe in each country also differs from others slightly. Kouglof can be said to be a cake or bread. Even if we're making a bread version of Kouglof, it borderlines on being cake-like. Its presentation is fine like cake. Its ingredients are basically those of a cake. The large amount of wine-marinated dried fruits add an extra dimension to the Kouglof that boasts light fluffy texture, strong buttery taste and rich aroma of the orange liqueur.

(makes 6 cakes, each 13 cm in diameter)

Wine-marinated dried fruit

30 g dried cranberries, 20%
30 g orange marmalade, 20%
30 g dried figs, 20%
30 g dried blueberries, 20%
30 g castor sugar, 20%
90 g Grand Marnier, 60%
30 g water, 20%
You may use rum instead of Grand Marnier if you prefer.

Autolyse dough

120 g French white flour Type 55, 80%
30 g bread flour, 20%
38 g castor sugar, 25%
45 g egg yolks, 30%
30 g whipping cream, 20%
30 g milk, 20%

Main dough

45 g sponge ferment, 30%
3 g salt, 2%
60 g butter, 40%

Other ingredients

45 g pistachios, 30%
almonds

Wine-marinated dried fruit

Put all ingredients into a pot. Heat over medium heat until it boils. Turn to low heat and simmer for 10 minutes. Transfer into a glass bottle. Let cool and store in the fridge for 2 to 5 days. Drain the dried fruits on a wire mesh for 1 hour before use.

Autolyse dough

Put all ingredients into the mixing bowl of a stand mixer. Mix until well combined.

Wrap the dough in the mixing bowl with cling film. Refrigerate at 4 to 6°C for 30 to 60 minutes.

Main dough

Let the butter warm up to room temperature (about 16 to 18°C).

In the mixing bowl of a stand mixer, put in sponge ferment. Tear the autolyse dough into pieces and put them in. Add salt. Beat with dough hooks over low spend until well mixed. Then turn to high heat and beat until smooth. Add half of the butter. Beat over medium speed until well mixed. Add the rest of the butter. Beat over medium speed until soft and well combined.

Add pistachios and drained dried fruits. Beat over medium heat until well mixed. Check the dough with windowpane test. It's done if you can stretch the dough into a thin membrane.

The dough should be at 22 to 23°C when the mixing is done.

Roll the dough into a ball. Put it into a plastic box.

First rise

Leave the dough at 28°C for about 3 hours until it double in size.

Adjust the rising time according to the condition.

Punch down / Cooling down

Press the dough to drive the air out. Put it into a baking tray lined with baking paper. Wrap it in cling film. Put in a freezer at -10°C and leave it there for at least 3 hours (up to 3 days).

Make sure you thaw the frozen dough in a fridge at 4 to 6°C before use.

Dividing / Resting

Cut the dough into 6 equal pieces with a scraper. Press each piece to drive the air out. Shape them round. Put them into a baking tray lined with baking paper. Wrap them in cling film. Refrigerate at 4 to 6°C to rest for 20 to 30 minutes.

Shaping

Grease the kouglof tins with butter lightly. Put the almonds on the bottom and set aside. Press the dough with your hands to drive the air out. Shape them round again. Grease your finger with a little butter. Poke through the dough at the centre. Put the dough into the kouglof tin with the seam facing up. Press the dough firmly.

Second rise

Leave the kouglof at 28°C for 120 minutes.

Baking

Preheat an oven to 220°C. Put the kouglof in and turn the oven down to 200°C immediately. Bake for 25 to 30 minutes. Remove the kouglof when time is up. Turn them out of the tin and brush on warm sugar syrup immediately. Let cool on a wire rack.

Sugar syrup

Make the sugar syrup when the kouglof are baking.
35 g castor sugar
30 g water
15 g Grand Marnier

Put castor sugar and water into a pot. Bring to the boil over medium heat. Turn off the heat and add Grand Marnier.

Lemon curd

zest of one lemon
15 g lemon juice
20 g whole eggs
15 g castor sugar
30 g butter

Put castor sugar, lemon juice and lemon zest into a bowl. Mix well. Add whole eggs and beat with an electric mixer until well combined. Transfer into a pot and heat over a pot of simmering water while keep stirring with a rubber spatula until it reaches 82°C.

Let the butter warm up to room temperature. Put it into a food processor. Add 1/3 of the egg mixture at a time. Beat well after each addition. Beat until silky smooth. Transfer into a container and refrigerate for later use.

To serve, put the lemon curd into a piping bag. Pipe some lemon curd into the indentation at the centre of each kouglof.

Chocolate Kouglof

Rich chocolatey bread with a hint of caramel sweetness

(makes 4 cakes, each 13 cm in diameter)

Autolyse dough

160 g French white flour Type 55, 100%
32 g castor sugar, 20%
48 g egg yolks, 30%
32 g whipping cream, 20%
56 g milk, 35%

Main dough

48 g sponge ferment, 30%
13 g cocoa powder, 8%
3 g salt, 2%
64 g butter, 40%

Spread

80 g Dulce de leche

Chocolate dip

115 g 65% cocoa chocolate
100 g whipping cream
23 g glucose syrup
23 g butter
chopped pistachios

Autolyse dough

Put all ingredients into the mixing bowl of a stand mixer. Beat until well combined.

Wrap the dough in the bowl with cling film. Refrigerate at 4 to 6°C for 30 to 60 minutes.

The autolyse process lets the flour to combine with water fully for better links between gluten and starches. It shortens the time required to mix the dough while preventing the dough to heat up too much. If the dough is too hot, the butter will melt and be absorbed by the flour. The kouglof will turn out dryer in texture, instead of moist and fluffy.

Please refer to P.185 for the main dough, First rise, punch down / cooling down steps in the kouglof recipe.

Dividing / Resting

Divide the dough into four equal pieces with a scraper. Press each piece with your hands to drive the air out. Shape them round. Put them into a baking tray lined with baking paper. Wrap in cling film and refrigerate at 4 to 6°C to rest for 20 to 30 minutes.

Shaping

Grease the Kouglof tin lightly with butter.

Press each piece of dough with your hands to drive the air out. Roll it out into a rectangle. Spread about 20 g of Dulce de leche over the dough with a palette knife, leaving 1 cm on all four edges uncovered. Roll the dough from the top down. Pinch the seam to secure well and keep the seam side facing down. Join the two ends of the dough to form a circle. Press the joining point to secure. Press the dough into the Kouglof tin with the seam facing down.

Second rise

Leave the Kouglofs to rise at 28°C for 120 minutes.

Baking

Preheat an oven to 220°C. Put in the Kouglofs and turn the oven down to 200°C immediately. Bake for 25 to 30 minutes. Remove the Kouglofs from the oven and turn them out of the tins. Let cool on wire rack.

Chocolate dip

Let the butter warm up to room temperature.

Put whipping cream and glucose syrup into a pot. Heat over medium heat until it boils. Pour the hot mixture into the chocolate. Leave it for 2 to 3 minutes. Beat with an electric mixer until well combined. Stir in butter. Dip the top half of the chocolate Kouglof into the chocolate sauce. Sprinkle with chopped pistachios.

Maple Cinnamon Apple Buns

Comforting cinnamon warmth with every bite

The spiciness of cinnamon and nutty sweetness of maple syrup are the perfect match with apples. The brioche dough almost has a cake-like texture and these buns are like a dessert course on their own. In one of those lazy wintry afternoon slouching on a couch, I always have a cup of strong tea and my maple cinnamon apple buns on the end table. With my favourite book on one hand and a bun on the other, I don't mind the long winter days at all.

(makes 4 buns, 13 cm in diameter each)

Cinnamon sugar (mixed well)

6 g ground cinnamon
40 g castor sugar

Maple cinnamon apple

18 g maple syrup
8 g butter
15 g castor sugar
150 g red delicious apples
20 g raisins
1.5 g ground cinnamon

Rinse and slice the apples. Put maple syrup, butter and castor sugar into a pot. Heat over medium heat until it boils. Turn to low heat and add sliced apples and raisins. Cook until apples are soggy. Turn off the heat and stir in ground cinnamon. Let cool.

Preferment

52 g French white flour Type 55, 20%
7 g sponge ferment, 2.5%
40 g milk, 15%

Soak the sponge ferment in milk for 2 to 3 minutes. Stir well. Add flour and mix again. Keep it at 28°C for 3 hours. Then put in a fridge and let it ferment for 18 to 24 hours at 4 to 6°C.

Autolyse dough

78 g French white flour Type 55, 30%
130 g bread flour, 50%
26 g castor sugar, 10%
78 g egg yolk, 30%
52 g whipping cream, 20%
52 g butter, 20%

Let butter warm up to room temperature.

Put all ingredients into a stand mixer. Mix until well combined.

The dough should be at 21 to 22°C when mixing is done.

Put the dough into a plastic box lined with baking paper. Refrigerate at 4 to 6°C for 18 to 24 hours.

The autolyse process lets the flour to combine with water fully for better links between gluten and starches. It shortens the time required to mix the dough while preventing the dough to heat up too much. If the dough is too hot, the butter will melt and be absorbed by the flour. The bun will turn out dryer in texture, instead of moist and fluffy.

Main dough

72 g sponge ferment, 27.5%
5 g salt, 2%
104 g butter, 40%
26 g milk, 10%

Let the butter warm up to room temperature (about 16 to 18°C).

Put preferment into the mixing bowl of the stand mixer. Tear the autolyse dough into pieces and put them into the same bowl. Add sponge ferment, salt and milk. Beat with dough hooks attached over low speed. Then turn to high speed and beat until smooth. Add half of the butter. Beat over medium speed to mix well. Put in the rest of the butter and beat with medium speed until soft and well combined. Check the dough with windowpane test. It's done if you can stretch the dough into a thin translucent membrane.

The dough should be at 22 to 23°C when the mixing is done.

Roll the dough into a ball. Put into a plastic tray.

Punch down / Cooling down

Press the dough with your hands to drive the air out. Put it on a baking tray lined with baking paper. Wrap in cling film. Keep in a freezer at -10°C for at least 3 hours (or up to 3 days).

Make sure you thaw the frozen dough in a fridge at 4-6°C before use.

Dividing / Resting

Cut the dough into 4 equal pieces with a scraper. Press each piece with your hands to drive the air out. Shape them round. Put them into a baking tray lined with baking paper. Wrap in cling film. Refrigerate at 4 to 6°C for 20 to 30 minutes.

Shaping

Grease the bread tin lightly with butter. Press the dough to drive the air out. Roll it into a rectangle. Sprinkle 12 g of cinnamon sugar evenly on the dough, leaving 1 cm on the left, right and top edges uncovered. Roll the dough from the top down and pinch the seal the seam. Put it on the counter with the seam facing down. Cut into 10 equal pieces. Put the dough pieces into the bread tin alternately with maple cinnamon apple.

Second rise

Leave the buns at 28°C for 120 minutes.

Baking

Preheat an oven to 220°C. Put in the buns. Turn the oven down to 200°C immediately. Bake for 25 to 30 minutes. Remove the buns from oven when the time is up. Turn them out of the tins and let cool on wire rack.

天然酵母麵包
WILD YEAST **BREAD MAKING**

作者 Author
Reko Sham

策劃 / 編輯 Project Editor
Catherine Tam

攝影 Photographer
Reko Sham

美術設計 Design
Charlotte Chau

出版者 Publisher
Forms Kitchen
香港鰂魚涌英皇道 1065 號 Room 1305, Eastern Centre, 1065 King's Road,
東達中心 1305 室 Quarry Bay, Hong Kong
電話 Tel: 2564 7511
傳真 Fax: 2565 5539
電郵 Email: info@wanlibk.com
網址 Web Site: http://www.formspub.com
http://www.facebook.com/formspub

瀏覽網站

會員申請

發行者 Distributor
香港聯合書刊物流有限公司 SUP Publishing Logistics (HK) Ltd.
香港新界大埔汀麗路 36 號 3/F., C&C Building, 36 Ting Lai Road,
中華商務印刷大厦 3 字樓 Tai Po, N.T., Hong Kong
電話 Tel: 2150 2100
傳真 Fax: 2407 3062
電郵 Email: info@suplogistics.com.hk

承印者 Printer
中華商務彩色印刷有限公司 C & C Offset Printing Co., Ltd.

出版日期 Publishing Date
二零一六年一月第一次印刷 First print in Jan 2016

Published in Hong Kong by Forms Kitchen,
a division of Wan Li Book Company Limited.
ISBN 978-988-8295-27-2